Saving
the
Wild
South

Saving the Wild South

THE FIGHT FOR
NATIVE PLANTS
ON THE BRINK OF
EXTINCTION

Georgann Eubanks

PHOTOGRAPHS BY DONNA CAMPBELL

THE UNIVERSITY OF NORTH CAROLINA PRESS

Chapel Hill

This book was published with the assistance of the BLYTHE FAMILY FUND *of the University of North Carolina Press.*

Designed by Richard Hendel
Set in Utopia and Klavika by Tseng Information Systems, Inc.
Manufactured in the United States of America

The University of North Carolina Press has been a
member of the Green Press Initiative since 2003.

All photographs, except those otherwise noted, by Donna Campbell.

Cover illustration: Cahaba lilies in Hatchett Creek in Alabama.
Photo by Elmore DeMott, www.elmoredemott.com.

Library of Congress Cataloging-in-Publication Data
Names: Eubanks, Georgann, author.
Title: Saving the wild South : the fight for native plants on the brink of
extinction / Georgann Eubanks ; photographs by Donna Campbell.
Description: Chapel Hill : The University of North Carolina Press, 2021. |
Includes bibliographical references and index.
Identifiers: LCCN 2021012150 | ISBN 9781469664903 (pbk. ; alk. paper) |
ISBN 9781469664910 (ebook)
Subjects: LCSH: Endangered plants—Southern States. | Rare plants—
Southern States. | Plant conservation—Southern States.
Classification: LCC QK86.U6 E53 2021 | DDC 581.680975—dc23
LC record available at https://lccn.loc.gov/2021012150

For my brother G. Ray Eubanks
explorer, mentor, scout

We could never have loved the earth so well
if we had had no childhood in it, if it were not
the earth where the same flowers come up again
every spring that we used to gather with our
tiny fingers. — George Eliot, *The Mill on the Floss*

When we speak about the conservation of nature,
we are really talking about a desire to conserve our
own human identity: the parts of us that are beautiful
and free and holy, those that we want to carry with
us into the future. — Nathaniel Rich, "Climate Change
and the Savage Human Future," *New York Times
Magazine,* November 14, 2018

Contents

Figures

Saving the Wild South

In the United States, the Endangered Species Act is our principal legal tool for protecting plants that are on a trajectory toward extinction. Although the U.S. Fish & Wildlife Service administers the Endangered Species Act for plants, no Federal agency, including the USDA Forest Service, can legally fund, permit, or carry out any activity that is likely to jeopardize the survival of a listed species. Once a species is listed under the Act, a recovery plan is developed that spells out actions each conservation partner can do to recover the species.

—United States Forest Service

Introduction

We live in a moment when the southeastern United States, one of the most biodiverse regions on the planet, is on the edge of redefinition. In only a few decades, we stand to lose a number of the natural features that have been a source of our identity and good fortune as a region and a people. Our lush forests, the sight and scent of blooming trillium and mountain laurel, the coastal azaleas, wax myrtles, and yaupon holly, the glory of summer's dense greening, and the magic of fall's deciduous palette are all at risk.

Already, many of the South's most unusual wild plants have been reduced to small, disconnected patches of land where they can thrive undisturbed. With the effects of global temperature change and rising sea levels, they may yet be lost. This book tells the stories of a dozen such plants, already recognized as threatened or endangered—plants that are highly prized by those who study and care for them.

But why should we care? Besides their connection to our common identity as residents of the South, they are links in a chain, interdependent species that rely on other plants and animals for their well-being. Losing one is like pulling a coin out of a stack. Eventually the pillar topples.

In *Smithsonian Magazine*, Bill Finch, the executive director of the Mobile [Alabama] Botanical Gardens, writes: "Conservation biologists are taking a closer look at the natural riches of the Southeastern United States, and they describe it as one of the world's biodiversity hotspots, in many ways on par with tropical forests." The distinguished Harvard scientist E. O. Wilson, an Alabama native and among the first to use the term "biodiversity," suggests that we need to invest more resources as a nation in preserving this region's rare plants.

This book invites readers—in an intimate and accessible way—to learn more about the extraordinary natural treasures of Alabama, Florida, Georgia, North Carolina, South Carolina, and Tennessee. Photographer Donna Campbell and I met with botanists, conservationists, botanical garden professionals, citizen scientists, and other advocates who study and protect native plants in the wild, and we discovered a range of destinations that readers can visit where endangered specimens have been placed in conservation. Seeing these ephemeral treasures up close forced us to consider our role in their future.

In obvious and accelerating ways, human beings have created conditions on the planet that have the potential to destroy whole environmental systems. Ice caps are melting to the far north and south, sea levels are rising all around us, hurricanes are strengthening and redrawing our shorelines, droughts are more frequent and prolonged, and wildfires are more vicious and relentless. Our collective sense of safety and our capacity to protect ourselves from the ravages of wind, water, and fire are now in doubt, if not already gone. Southerners along the Atlantic and Gulf Coasts, in particular, who have been forced to rebuild time and again after more frequent and devastating hurricane damage are finally seeing the folly of big development on our shorelines. In September 2018 multiple weather experts predicted that Hurricane Florence would set a new record for East Coast rainfall. The storm dumped the equivalent of all the water in the Chesapeake Bay, mostly on North Carolina.

No surprise, then, that insurance companies are either retreating from protecting coastal properties or simply raising policy prices to impossible levels, and the US government's federal flood insurance program is insolvent and headed toward bankruptcy.

Despite our enormous human capacity for denial, it does not take much vision to see that life as we have come to know it—depending on fossil fuels and free-flowing herbicides and pesticides, on single-use plastics and nonrenewable resources—is unsustainable. Meanwhile, the historic natural landmarks and other beloved assets in the countryside are disappearing in such a way that our identities as a diverse collection of peoples and cultures in the South are literally being erased.

I asked experts to point me in the direction of endangered plants with interesting stories, plants that might teach us something in their struggles to survive human encroachment. The southeastern United States is home to the longleaf pine savannah, "once the dominant tree of 60 percent of the land from the Carolinas to eastern Texas," E. O. Wilson writes. "Pitcher-plant bogs scattered through the longleaf woodland are in turn among the richest in the world, comprising up to fifteen slender-stemmed species into a square meter." We will visit these fascinating carnivorous plants and consider the loss of our longleaf woodlands in favor of the ubiquitous loblolly pine plantations, still dominant and reseeding vigorously in most places where they replaced the longleaf pines. Now, though, paper mills are disappearing dramatically in the region as cheaper sources over-seas have made our homegrown pulpwood less competitive at the same time that digital communication has brought about a decline in the use of paper.

Certain plants and trees of the wild South are iconic because this region was a destination for many of the earliest academic botanists who worked on the continent, explorers who collected and named tens of thousands of extraordinary plant species in their travels. But it is precisely because of the region's dense biodiversity that there is also a higher number of species at risk here. Still, it is news to most people that biologists working over the centuries have discovered and recorded perhaps only 20 percent of all species on the planet—plant, animal, and other organisms. "Flower-ing plants come in with about two hundred seventy thousand species known and as many as eighty thousand awaiting discovery," E. O. Wilson writes. "Scientists working on biodiversity are in a race to find as many of the surviving species as possible in each assemblage ... before they vanish and thus are not only overlooked, but not to be known." In Alabama, for example, I visited the ruins of an antebellum industrial site where sixty-one rare plant species flourish—eight of which were discovered only in the 1990s by Nature Conservancy botanists.

The special relationship of plants and trees to people in the South began centuries before glossy magazine photos of magnolias, camellias, and azaleas created a certain aristocratic image of the region. The Native American tribes that flourished here before European explorers arrived were the first to cherish and depend on native plants for food, medicine, religious ritual, and functional items such as river cane baskets and mats. Today only 2 percent of the river cane (*Arundinaria gigantea*), once so pro-lific, still exists in the region, making it difficult for the Cherokee people to secure materials for their celebrated basket- and mat-making. Choc-taw and Creek (presently known as Muscogee) Indians in Georgia and Alabama also made flutes, arrows, and baskets from this plant. Although the US Fish and Wildlife Service has not listed river cane as endangered, thickets of cane provide critical habitat for the Swainson's warbler and six butterfly species on the government's endangered list.

Another plant, Miccosukee gooseberry (*Ribes echinellum*), is an ancient shrub, named for a small Indian tribe that once lived in the Florida Pan-handle. The plant now grows only in Jefferson County, Florida, on private land, and on public land in South Carolina's least populous county. This special fruit-bearing species began its slow decline centuries ago, possibly with the extinction of the ancient mastodons that grazed on branches and leaves of overstory trees, allowing crucial sunlight to stream onto lower shrubs in its habitat. More profoundly, the changes in land use when Europeans arrived on the continent eventually led to a lack of genetic diversity among the few remaining specimens and is a major factor in the plant's critical endangerment. Is the Miccosukee gooseberry a serious loss, or is its demise a natural consequence of the evolving landscape? What to protect and what to let go? Botanists have differing opinions.

Seeing these plants in the wild brings it all home. The South is where I was raised and have lived all my life. My father's people were from Alabama; my mother's family was from Georgia. As a child I couldn't wait to get to the white sand beaches of Florida or South Carolina on summer vacations. My family also took fall and spring trips to the Smokies in Tennessee and the Blue Ridge in North Carolina to hike in the mountains and to breathe in the tangy scent of Fraser fir, also now endangered in the wild. We'd delight in the profusion of wildflowers that my grandmother Stella could instantly identify with both their Latin and common English names. The love of place, plants, and trees has become the part of my heritage that I most treasure.

Memories of childhood escapades in natural settings flooded back as Donna, a North Carolina native, and I traveled to research this book. Seeing the radical changes in once familiar landscapes made us ache for the losses.

I found myself wondering at the confidence people once had in "scientific" solutions to natural "problems," such as insect pests. In 1960, when I was six, it seemed perfectly safe to ride my bike in a swarm of kids rushing headlong into the acrid chemical fog behind the mosquito truck that rumbled through St. Andrews State Park in Panama City, Florida, while my parents pitched a tent and fired up the grill for supper. Nor did it occur to me to be afraid when, in the summer before third grade, retrofitted DC3s from World War II flew low over my grandparents' garden and fish pond in Atlanta, dropping fire-ant killer—a poison that had been, of all things, soaked into Georgia grits for broadcast over residential areas in the suburbs. We were told that the enemy in this case was the stinging red fire ant. The insects had come as stowaways on cargo ships from South America and landed in Mobile, Alabama. They migrated across the South, building huge mounds in cleared fields and attacking any creature—including kids—who disturbed their nests, though I had never seen them where I lived.

The chemical that most likely was dropped on our heads that summer was Mirex—introduced as a putatively benign substitute for hepta-chlor. Rachel Carson had made heptachlor notorious in *Silent Spring*, her 1962 book about the unacknowledged and untested hazards of synthetic chemicals applied to soil, water, and air, and thus to animals and people. By the end of the 1960s, however, Mirex residues from this airborne cam-paign were detected in nontarget organisms. The chemical was also found toxic to estuarine life and was declared to be, like heptachlor, a potential carcinogen. By the time Mirex was taken off the market, in 1978, I had graduated from college.

I can still hear those grits faintly pelting the oak, poplar, and sweet gum leaves in my grandparents' yard, with a sound like falling sleet amid the summer birdsong. The planes flew thrillingly close to the treetops and came twice a day. Bored on those summer mornings, I would listen for the planes. At the first rattle of dishes in the cupboard, I'd run out of the house and down to a clearing in the yard to see them pass over, my mouth agape.

The extent of human destruction in my home territory over my life-time is vast. Commercial development has taken away the houses that my grandparents, aunts, uncles, and parents built in the 1940s and 1950s. The creeks fed by springs and flanked with wild fern and ginger root—the places where I played as a child—have been diverted into buried culvert pipes. The modest ranch houses on five-acre lots where my people put down roots have been bulldozed and reconfigured into dense, high-end developments often landscaped to golf course standards—meaning grass lawns peppered with pesticides and herbicides—and tended by absent property managers whose services are paid out of homeowners' dues.

As I write, the poisons flowing into our streams, rivers, groundwater, plants, and bodies—once limited for a time by law—have ballooned again with the loosening of regulations in Washington, DC. The escalating loss of woodlands to roads and rights of way, the conversion of southern prairies, pine barrens, and forests to farmland, and the wholesale development and redevelopment of cities and suburbs have left conservation botanists working in strange checkerboard patterns of small, disjunct patches of land. A satellite map released by University of Cincinnati geography re-searchers in late 2018 showed that nearly a quarter of the earth's habitable surface changed between 1992 and 2015, primarily from forests to agricul-ture, from grasslands to deserts, and from wetlands to urban concrete. In 2019 alone, the warming of ocean temperatures reduced scallop harvests in the Atlantic. At the same time, immense floods along the Mississippi River unleashed so much fresh water into the Gulf of Mexico that the oys-ter yield there was falling.

Adding to the menace of climate change, many plants' native habitats are actually "moving" north as global temperatures rise, making safe haven quite out of reach for certain species unless there is intentional human intervention. The notion of moving or replanting certain trees, shrubs, and native flowers in a new landscape that's better suited to their historic environmental needs has created a passionate debate among botanists. Our original understanding of the term "native" is being altered by climate on a much shorter timetable than previously expected. The assisted migration of an endangered species—*Torreya taxifolia* also called the Florida nutmeg or gopherwood tree—is part of a still-unfolding story that is discussed in chapter 2.

Given our many human intrusions on the natural world, some environmentalists argue that using the terms "wild" or "wilderness" is inaccurate and obsolete. As a remedy, E. O. Wilson proposed setting aside half of the earth's wild areas to let nature reclaim them. Younger bioethicists argue that wilderness is unrecoverable. Nonintervention cannot ensure long-term biodiversity, as many scientists and conservationists believed in the twentieth century. Emma Marris, author of *Rambunctious Garden: Saving Nature in a Post-wild World,* says that no matter where we go, the profoundly increased presence of carbon dioxide is in every breath we take. Trying to return any given region on earth to an earlier baseline, she argues, is folly. "The cult of pristine wilderness is a cultural construction, and a relatively new one. It was born, like so many new creeds, in America," Marris adds.

Environmentalism, too, is a relatively new movement, launched in the United States in part by the establishment of Yellowstone National Park in 1872. This legal precedent for placing natural preserves under federal jurisdiction was followed by the Antiquities Act of 1906, which guaranteed the preservation of "historic landmarks, historic and prehistoric structures, and other objects of historic or scientific interest" on lands owned or controlled by the United States.

The establishment of the National Park Service followed in 1916 and eventually led to the passage in 1966 of the Endangered Species Preservation Act, which initially focused only on protecting endangered animals. In 1973 the act was expanded to include endangered and threatened plants. By 1975 a study by the Smithsonian Institution recommended more than three thousand plant species in the United States for possible listing as threatened or endangered. The original privileging of fauna over flora also tells us something about the implicit bias of human beings toward animals and a spreading condition among young people that botanists call "plant blindness."

Setting aside national parklands in the western states to preserve a bit of the frontier mythology fundamental to the Euro-American narrative also unintentionally emphasized the idea of human separation from and superiority to nature, in much the same way that traditional museums distance us from our past by placing historic artifacts under glass. Saving Yellowstone and other national monuments and parklands came about as a response to the rapid and rapacious industrialization that was sweeping over the eastern half of the country in the late nineteenth century. Roderick Fraser Nash suggested in his 1967 discussion of the origins of the conservation movement, *Wilderness and the American Mind*: "One man's wilderness may be another's roadside picnic ground." In other words, the idea of wilderness is relative to the observer's experience. Even Thoreau, meditating on the natural world at his much-romanticized Walden Pond, was only a short walk away from the bustle of Concord, Massachusetts.

And what do children today know of plants and wilderness? Richard Louv's *Last Child in the Woods*, published in 2005, warned parents of the ill effects of what the author called "nature deficit disorder," created in part by parents afraid to let their children roam in the woods to play as earlier generations did. It is news to no one that this condition has been amplified in the years since by the seduction and dominance of digital devices in our hands.

The conflicted sentiments that human beings hold toward nature—as something to be feared or revered, dominated or discounted—seem to have amplified our sense of estrangement from the wild rather than solidifying our understanding of ourselves as integral to it.

In this century we are an undeniable part of nature, more so than ever before, thanks to our carbon emissions. In *Falter: Has the Human Game Begun to Play Itself Out?*, the environmental journalist Bill McKibben writes, "Human beings are now a geological force. In fact, we are one of the half-dozen or so largest geological forces to punctuate the billions of years of earth's history."

We must own this truth now and reckon with it. As climate change and rising seas force the migration of peoples to higher ground, the next generation must adapt to the consequences and act on behalf of all flora and fauna, human beings included. How shall we prepare our children and grandchildren to comprehend these changes and losses and imagine new possibilities?

This little book begins humbly with a few basic examples from one small region known for its biodiversity. What can we learn, along with our progeny, by getting acquainted with a dozen plant species on the brink of extinction in the South? What do we find when we stop long enough to visit these plants and understand the factors that have contributed to their near-demise?

The plants and trees discussed here, along with their human caretakers and defenders, stand for thousands more. This book examines the range of threats they face, including underfunding of the work of scientists and the activists who call attention to their situation. It offers readers a chance to visit sites where they can easily and safely see these plants on field trips as part of a vacation or daylong excursion. These destinations include botanical gardens, public parks, conservation sites with occasional public visiting days, and other unusual touring opportunities. The book does not reveal any locations in the wild where fragile plants are being guarded and studied and would be further endangered by visitors.

My hope is that we might find in ourselves (and instill in coming generations) a deeper and more accurate sense both of the peril and amazement that's at stake and cultivate new habits in our lives that could make a difference to our native botanical treasures in this distinct geography we call the US South.

in the Piedmont after classes ended for the school year. They then hired a horse-drawn hack to take them from Salisbury to Lenoir in the Blue Ridge foothills, where a wagon service would carry them through Happy Valley and up the ridge to Blowing Rock.

Small and Heller, young Victorian-era men from modest backgrounds, were ambitious and single-minded. Clearly, they felt their own sap rising as they prepared to identify and collect specimens of wild plants known and unknown in these parts. Dressed in the formal clothing of the era, the two would travel that summer for weeks—mostly on foot, sometimes by horseback or buggy, and once on a small-gauge train—to explore the high country. They scaled the summit and grassy balds of Roan Mountain at the Tennessee border. They hiked to and from Grandfather and Table Rock Mountains in North Carolina—destinations I have come to love deeply over thirty years of my own explorations along the Blue Ridge Parkway and the trails that fan out from it.

The great Ice Age glaciers that covered North America stopped just short of this region, accounting for its exceptional biodiversity. Small's entertaining and sometimes tongue-in-cheek memoir of the 1891 expedition with Heller was published in the prestigious *Bulletin of the Torrey Botanical Club* of New York less than a year after their adventures. There Small elaborates on the hardships of plant hunting in the steep and sometimes impenetrable thickets of rhododendron and mountain laurel of the Blue Ridge. "It cannot be recorded here how many times we lost the way, how the horse gave out and walking had to be resorted to, the accident that happened to the rations, and other mishaps," he wrote.

The young men carried their survival supplies along with an assortment of cylindrical tubes or portfolio boxes, known as vasculums. Outfitted with shoulder straps, these tubes—often made of decorated tin or a stiff, suitcase-like cardboard—were designed to store plant samples safely until they could be properly dried and pressed as specimens suitable for further study. Dried out and mostly flat, the plants would then be packed for the return to Pennsylvania.

Once back in a collector's lab or accessioned into a museum collection, the plants would be carefully mounted on archival sheets of thick paper with notations on their provenance, the collector's name, the date of acquisition, and the Latin or Greek genus and species names. If the plant was previously unknown and judged to be a new find, it could receive a new species designation—sometimes the Latinized version of the name of the botanist who discovered it. Over time, these sheets might be given additional handwritten notes and observations from other collectors who studied the specimens and verified them. This oldfangled practice of clipping, drying, pressing, and preserving plant specimens is still the way taxonomy is practiced today, though new technologies have been added to species determination.

Small and Heller were intentionally following in the footsteps of several famous botanists who had earlier documented thousands of the region's wildflowers and trees, creating a list that is still, in our time, far from comprehensive. Among the first British explorers to come to the South were the Philadelphia-based naturalist John Bartram and his son, William, who became a skilled botanical illustrator. William navigated the region in 1775 and published his findings on the botanical riches of the South, giving a nod to the broad knowledge and experiences of the Indigenous people who sometimes showed him what they knew of native plants and their useful medicinal and gustatory properties.

Later, the French government sent the botanist André Michaux, born in 1746, to collect live plants in North America and ship them back to his home country. Michaux followed the route of William Bartram through Georgia and on to North Carolina and Tennessee; he was the first academically trained botanist to scout the Volunteer State. He established a nursery in Charleston, South Carolina, in 1787 and stocked it with 2,500 young trees, shrubs, and other plants—many from the Appalachian Mountains. His last expedition to the Blue Ridge, from 1794 to 1795, included the summits that Small and Heller would later climb. Before his death in 1802, Michaux had sent sixty thousand woody plants and forty boxes of seeds back to France from his North American travels. Today, if you keep a sharp eye out, you'll see roadside historical markers that document his path across several states.

As the eighteenth century ended, European botanists, including Michaux's son, François André, continued their explorations of the South. Not to be outdone by the Europeans either politically or scientifically, Thomas Jefferson, an enthusiastic horticulturist, asked Congress for the funds to broaden the young nation's environmental and economic explorations. In 1803, the year after Michaux's death, Jefferson—who had already imported many species of plants and seeds from France to his Virginia estate—tapped Meriwether Lewis and William Clark to lead a team of US explorers as far as they could travel westward on the continent.

Michaux's work would inspire Harvard University professor Asa Gray (1810–88), an American-born physician who first served as an assistant to the noted New York City botanist and medical doctor John Torrey (1796–

1873). Gray was a brilliant scientist who completed his medical degree before the age of twenty-one and would sustain a lifelong friendship with his mentor, Torrey. In 1836 Gray published *The Elements of Botany*, a textbook surely studied by the young botanists Small and Heller when they were in college. Two years later, Gray and Torrey jointly launched the journal *Flora of North America*. The publication kept up, somewhat erratically, with ongoing developments in botany, now a burgeoning field on the continent.

Gray corresponded with Charles Darwin and was a respected advocate of Darwin's controversial theories. Though Gray was not known for his skills as a lecturer, his friendly writing helped nonbotanists, especially farmers, understand the field better. Over his lifetime, Gray collected more than two hundred thousand plants from North America. Gray's lily, with its stunning, reddish-orange trumpet flowers, which Gray discovered in 1840, still grows on Roan Mountain and a few other spots in Tennessee. It is now a highly vulnerable species prone to a fungus spread by human touch. The Asa Gray Herbarium, at Harvard, honors Gray's contributions to the field.

John Torrey was also celebrated for his many years of botanical investigations. His colleagues launched the Torrey Botanical Club, which counted many amateur botanists and students among its early members. In 1867 Torrey became the club's first president. Now called the Torrey Botanical Society, it is the oldest such organization in the Americas. The group began publishing the *Bulletin of the Torrey Botanical Club* in 1870. Torrey would finish his career as a professor of chemistry and botany at Columbia College in New York and donate his personal collection of fifty thousand dried plant specimens to Columbia to create that institution's first herbarium. Of Gray and Torrey, the botanical historian Frederick Brendel waxed jubilant in 1879: "These two bright stars had already risen above the horizon of the botanical firmament, opening a new epoch in the history of American botany. The interest in the science of botany was now wide awake amongst the American public, and the Government bore its rich share of it, spending large sums for scientific purposes, by attaching scientific men to the nearly unbroken series of expeditions and surveys which were now undertaken."

Despite Brendel's optimistic outlook on government funding for botanical research, Arthur Heller and John Small did not come to the Blue Ridge with financial support for their investigations. They were compelled only by a passion for the natural world and the thrill of discovery. In remote North Carolina they often depended on the kindness of local hosts to feed them and provide shelter for the night. They ate what they could carry into the field or find in small communities along the way. They often despaired when afternoon thunderstorms blew up and ruined the samples they gathered beyond what would fit inside a vasculum. But clearly the young men were driven to the heady business of classifying and naming their botanical finds by properly identifying the genus and then determining whether the species was something previously seen elsewhere or whether it might bear traits that were unique. Small and Heller's facility with the Latin names of thousands of plants, as demonstrated in Small's journal of the trip, is dizzying. Despite Small's use of the passive voice in an attempt to be objective and impersonal, the personalities of these two adventurers shine through the narrative.

Out for a walk after setting up their headquarters in Blowing Rock—whence they planned to venture out in all directions through the summer—the botanists came upon the local Catawba rhododendron in full bloom with deep pink flowers. It was, Small wrote, "a great quantity of *R. catawbiense*, bushes ten feet or more in height, and between the two we almost lost our heads."

After a few days settling into their new quarters, the boys set out to hike to the nearly six-thousand-foot summit of Grandfather Mountain, so named for its unmistakable profile of a bearded man in repose, as if he were taking a nap. The June profusion of wildflowers and woody shrubs that bloomed among the rocks and ridges all the way to the windy summit awed the plant hunters, who took copious specimens.

Tired and hungry on the way back, they came upon "a good-sized bed of *Fragaria virginiana*, evidently uncultivated, but the fruit large and attractive looking," about five miles west of Blowing Rock. Small wrote, "Many dozens of them were collected, but the receptacles into which they found their way were not portfolios." The wild strawberries went straight into their mouths.

About a month into their adventure, Small and Heller took a break from the rigors of plant hunting. "On the 13th of July our headquarters were filled almost to overflowing by the arrival of a party of friends from Salisbury, among whom were six young ladies. To say that we had an enjoyable time—as far as it was in the power of two hard-working collectors—is a very temperate expression."

When the party was over, Small and Heller prepared to conquer Table Rock, a well-defined peak overlooking the wilderness of Linville Gorge. Table Rock would require a rugged journey of thirty-five miles and a steep ascent. (It is now possible to drive up the mountain on a narrow road, park, and hike only the last mile on foot.) On their long hike, in the vicinity of Wilson's Creek, the young men heard the unmistakable warning of a rattlesnake in the brush. They cut a wide berth. The encounter shook them, and soon the explorers realized they had another problem:

> Our supply of rations for the day was somewhat limited at the start, and it was lessened considerably when one of us discovered that his share had disappeared. Of course, he began to get very hungry as soon as the loss was noticed, and there was great lamentation when he pictured how some wandering dog, or, perchance, a bear, would feast on his chicken and biscuits. Finding at length that the outer man was unable to make much head way, it was decided to fill up the inner one at the first suitable place. Pie was demanded, but there was nothing of that description to be had. Cold potatoes and cold cornbread, the heaviness of which was enough to cause a man with a cast-iron digestive apparatus to turn pale, were set before us. We ate, and, to our astonishment, are still alive.

As they finally began climbing the dry and rocky trail toward the summit of Table Rock, they were surrounded by a cloud of small, pesky bees that flew into their hats and clothes. Two months later, Small would unwrap a batch of Table Rock plants they'd sealed up and shipped home. "One of the little insects walked out," he wrote.

The accumulation of plant specimens grew as the duo covered mountain after mountain, and the summer breezed along. Back at headquarters near the end of their adventures, Small and Heller were spooked again. They came upon "a brace of wildcats just after nightfall as we were climbing the slopes of Blowing Rock mountain." Then, Small explained, "upon close inspection [the wildcats] turned out to be a branch with two bunches of leaves on it."

These young men subjected themselves to strenuous challenges in the service of science. As a child I somehow believed that scientific inquiry such as this had already conquered the kingdom of animal, vegetable, and mineral. In grade school we were taught basic science via a grainy black-and-white television program shown once a week for half an hour while our regular classroom teacher got to stop talking for a while. She would sit at her desk eyeing us and sipping from a glass of iced tea while we trained our eyes on the same TV-on-wheels that we had stared at when John Glenn become the first American to orbit Earth, on 20 February 1962.

Our TV science teacher, Miss Annie Flanagan, generally confined her lessons to Earth and twanged along in her north Georgia accent, one day pulling out a basket of chicken eggs and opening them to reveal the embryos inside. In the last one she cracked open on that day's show, the creature inside (not a yolk) started moving and chirping. In another lesson she explained the difference between igneous, sedimentary, and metamorphic rocks. I still remember that she pronounced mica as "MY-cur" and Georgia as "GEORGE-er." Miss Annie was also missing part of a finger, which made her desperately interesting to the boys in my class, who would yell "Look!" every time she picked up one of her eggs or rocks for a demonstration. At least we were paying attention.

As an adult, I have come to appreciate that hunting natural objects in the wild requires the cultivation of a very specialized kind of attention. The minute details of leaf shape, of the play of sunlight filtering through the overstory, of land slope and soil composition, and of moisture—seep and drainage—are all critical. The competing demand to watch for snakes, poison ivy, and bears or, farther south, alligators is also essential. This ability to focus on the smallest details in a landscape is a handy skill to cultivate. But many young people today do not have eyes for the natural world. The inability to see the differences among plants, say botanists, has become a real threat to the value we need to place on the biodiversity and beauty around us.

The term "plant blindness" was coined by two botanists from the American South—Elizabeth Schussler, of the University of Tennessee, and the late James Wandersee, of Louisiana State University in Baton Rouge. Wandersee and another colleague, Renee Clary, described the condition: "Most people in developed nations tend to see plants as merely a green blurry backdrop for the animals and human-made objects that populate their visual field."

Wandersee began a movement among science instructors to beef up their teaching about plants to help students overcome the human tendency to prefer animals over plants. In a 2006 study the researchers surveyed high school students at graduation and found that "seventy-seven percent of those questioned had never grown a plant by themselves. Ten percent had never picked fruit from a tree, nor could they remember any children's story that featured a plant."

This plant blindness, educators believe, has led to a decline in funding for plant conservation initiatives and the study of botany itself as compared to animal-related studies. Academic departments of botany and horticulture are currently in decline across many universities, and not only in the United States.

Learning about Small and Heller's summer in the North Carolina mountains gave me some notion of the demands of botanizing, as it is called, but it could not fully prepare me to confront my own long-cultivated comforts—spending summers at a desk, cool and dry, traveling the world from an armchair with only the shine of a computer screen on my face. Getting out into the wild spring and wilder (and hotter) summer to study blooming things was quite another occupation. I, too, had forgotten the joys and demands of being out in solitary nature with so much to take in. I can hardly imagine how it must have felt to Small and Heller to explore the wild South when they did.

Preparing to leave the North Carolina mountains and head back to Pennsylvania, the young botanists ordered lumber and constructed shipping containers for their specimens. They had enough "to fill two baggage wagons," Small wrote. In the rising August heat, they decided to make one last climb along the ledges of Blowing Rock. They were hoping to spot the August-blooming *Liatris helleri*—a previously undocumented species of aster that Arthur Heller had first collected in the region the year before, naming it after himself. The federal government has declared Heller's blazing star—or Heller's gayfeather, as it is commonly known today—a threatened species. Found in five North Carolina counties and a few other spots in the Blue Ridge, it is protected on Grandfather Mountain: a preserve owned by the State of North Carolina and designated since 1992 a member of the United Nations international network of Biosphere Reserves. Other, more prolific cousins of this member of the aster family have been used to formulate drugs in the treatment of leukemia. However, the scarcity of Heller's species has meant that it has never been formally tested for its potential medicinal benefits.

Using the passive voice as usual, Small describes the pair's final adventure on the cliffs at the edge of Blowing Rock, on 8 August, without identifying which of the botanists had once again gotten into trouble:

Many of the ledges are only a few inches wide and are not by any
means safe places on which to ramble about, as the base of the cliff is
more than one hundred feet below. One of us has cause to remember
the place, for on that day his early career was almost ended. While
carefully picking his way along one of the narrow ledges ... his foot
slipped and over he went, turning somersaults, and desperately
clutching at anything that offered support. A narrow shelf and a
friendly bush finally stopped his descent, after he had fallen about

fifteen feet. Two badly damaged fingers and several minor bruises were, fortunately, the only results.

The flowers they collected on the rocks that day were not *Liatris helleri* but *Liatris graminifolia*, also known as grass-leaf blazing star, aptly named since the leaves are narrow and grasslike. It was too early, Small later wrote, for his companion's namesake flower, Heller's blazing star, to be blooming.

A few days later Small and Heller descended the Blue Ridge and made their way to one last destination for botanizing: the banks of the Yadkin River, near the geographic center of North Carolina at the edge of the Uwharrie Mountains. Though the Uwharries' ancient crenelated crests reach to heights barely a thousand feet, on a clear day at the top you can see the much taller Blue Ridge escarpment to the north and west. Nowadays, the Uwharries are the only designated wilderness in the Piedmont region of North Carolina. This woodland expanse bears the marks of earlier human beings in residence. Native American artifacts and remnants of settler cabins are tucked in the woods and have been the subject of a large archaeological excavation. Here, too, is the Yadkin River, which changes its name to Pee Dee where it intersects with the smaller Uwharrie River and then flows down into South Carolina.

Heller and Small focused their plant gathering on a section of the Yadkin known as the Narrows, more than one hundred miles downstream from where they had been near the river's headwaters, in Blowing Rock.

There they came upon a goldenrod—taller and with larger flower heads than any other specimens they had seen on their expeditions. Small made note of the unusual plant in his journal.

After publishing the results of that summer's odyssey in the *Bulletin of the Torrey Club*, Small was offered a fellowship to study for his master's degree at Columbia College. He would end up giving many of his specimens from North Carolina to the herbarium that John Torrey had launched in New York City.

When he finished his master's degree in 1894, Small made another visit to the Yadkin Valley and came upon the tall goldenrod once again. Convinced it was a unique species, he gave it a Latin name: *Solidago plumosa*, for its plumage—much larger than that of its nearest cousin, *Solidago purshii*. Writing once again for the Torrey Botanical Club, Small offered these notes:

> In the latter part of August 1894, I was surprised to find handsome specimens growing in crevices of the rocks at the bottom of the cañon

Two rare species discovered in the late nineteenth century in North Carolina by John K. Small and Arthur Heller: (left) the Yadkin River goldenrod (Solidago plumosa), photographed by Julie Tuttle in Stanly County, North Carolina, and (right) Heller's blazing star (Liatris helleri), photographed in western North Carolina by Tom Earnhardt. Used by permission of the photographers.

at the falls of the Yadkin river, and at the Narrows some miles above the falls in Middle North Carolina. The plants differ from specimens in the northern localities in their greater size and glabrous achenes [smooth seeds]. There is an abundant growth during the last part of August and the first weeks in September, but at other seasons hardly a vestige of the species can be found. The average height of the plants is about three feet; a few specimens attained a height of four feet.

Fast-forward one hundred years, to August 1994. Alan Weakley, an up-and-coming conservation biologist, was collecting data along the Yadkin River for the Carolina Vegetation Survey, an official census of rare native plants funded by the State of North Carolina. Weakley, then in his thirties, was in the last month of a job he'd had for ten years. "I was driving around the state on back roads," he explains, "looking for high-quality ecosystems and documenting rare species that might be worthy of conservation."

Weakley's official title was state botanist and ecologist in the North

Carolina Natural Heritage Program—a position that a Raleigh newspaper once listed as among the ten best jobs in the region. Sitting now in the library of the North Carolina Botanical Garden, where he is director of the herbarium and a much-in-demand speaker on native plants of the South, Weakley laughs. "I guess that was true," he says. "It was always an adventure to be out scouting the woods."

At about three o'clock on that bright August afternoon, Weakley and his field assistant finished their survey for the day. They looked once more at the collection of aerial photos they were using to find their way around the forest along the Yadkin River. "This was long before Google Maps," Weakley says with a grin, his blue eyes bright behind old-fashioned, frameless spectacles. "Back then," he goes on, "field biologists struggled just to get good color photos taken from airplanes. And even if you did have the images, you had to learn how to read and interpret them."

In one aerial photo, Weakley noticed an area beside the river where rock outcroppings were visible, just below the historic Narrows Dam, built by Alcoa Aluminum as a part of a smelting operation they had taken over from a French competitor in 1915. The rocks put Weakley in mind of an entry in a scientific journal from a century before that John K. Small, a young botanist just getting his start, had written. Weakley had been amused by Small's lighthearted approach to plant hunting.

When he read Small's narrative, Weakley had tried to cross-reference the Yadkin River goldenrod in the 1968 *Manual of the Vascular Flora of the Carolinas*—still a popular reference today. It was not mentioned. But Weakley and a colleague—Steve Leonard, a former curator of the North Carolina Herbarium—had only recently had a conversation about the disappearance of the Yadkin River goldenrod. They both had studied Small's notes that explained how he first saw the plant on a certain stretch in the Narrows Canyon of the Yadkin. The plant was described as coming up after heavy rains had scoured the dark, igneous bedrock along the shoreline, removing silt and exposing the rocks to sun. That goldenrod found purchase in crevasses between the rocks, Weakley remembered.

Perhaps, he thought, this plant with such an extremely limited range had been driven to extinction when Alcoa dammed the Narrows in 1917 to create Badin Lake, where speedboats and Jet Skis now rocket through the waters and break the quiet of the surrounding forest. Alcoa had also built a hulking hydroelectric generator beside the water. The millworkers used to live in a planned community of rowhouses near the lake, originally laid out by the French proprietors of the smelting operation. Badin was a company town, and Alcoa taught the local residents how to manufacture ingots of pig aluminum. The operation grew. The company then constructed a second dam in 1919 and, below that, a third dam in 1928 that created Lake Tillery.

On that humid August afternoon in 1994, Weakley and his assistant, Tom Phillipi, drove down below the first dam and scrambled through the bush to reach the water's edge.

"And there it was," Weakley says. "I know my goldenrods. It was the one John K. Small first described." Weakley is invigorated by the memory. "That day was one of my top ten," he says, leaning forward. He found the goldenrod where the rocks get a scouring when water is released from the dam.

As it turned out, Weakley wasn't the only botanist who had found the plants in that same twenty-four-hour period in 1994. The very next day, Steve Leonard, with whom Weakley had recently discussed the lost goldenrod, went to the area and found the plants himself. It didn't take long for the two scientists to confer and celebrate their mutual finding, exactly one hundred years after Small had been collecting his samples of *Solidago plumosa*.

After the summer of 1891, Arthur Heller returned to Franklin and Marshall University and would earn his master's degree in botany in 1897. Following a teaching stint at the University of Minnesota, he was engaged by the New York Botanical Garden to work on the Vanderbilt expedition to Puerto Rico, where he collected amazing tropical specimens. Heller eventually moved to California. He stayed in contact with his old friend, Small; continued botanizing; and taught at the college and high school levels over the remainder of his career. Today Heller's earliest collection of more than ten thousand specimens is housed at the Brooklyn Botanic Garden. His collection of plants from Puerto Rico was given to the University of Washington in Seattle, where they are prized for their usefulness in the study of endangered species.

Following his degree program, Small continued working in the Columbia College Herbarium until the school decided to transfer their specimen collections to the up-and-coming New York Botanical Garden (NYBG), founded in 1891, the same year that Small and Heller had been in Blowing Rock. Small went to work for the NYBG, taking along the Columbia specimens and becoming the institution's first curator of museums. He was promoted to chief curator in 1906 and then chief research associate and curator in 1934.

As his career matured, Small turned his attention to the flora of Florida,

Throughout his distinguished career, Alan Weakley, director of the University of North Carolina Herbarium, has continued to collect plant specimens on rock outcroppings at the edge of the Yadkin River near Badin, North Carolina. Below this site, the Yadkin is called the Pee Dee and flows into South Carolina. Used by permission of the photographer, Julie Tuttle.

and his plant-hunting excursions were conducted under the auspices of the NYBG, well on its way in the first decades of the twentieth century to becoming one of the top botanical organizations on the planet.

Small married and raised four children with his wife, Elizabeth. (He also played flute in the New York Philharmonic Orchestra.) He would often bring along his wife, two sons, and two daughters on extensive plant-hunting jaunts in Florida. By this time, bold new pathways were being cut into formerly inaccessible Florida hammock—a situation that gave Small fresh opportunities for discovery, along with a profound sense of loss. Wild Florida was not so wild anymore. Yet, as we now know, the human impact on the wild South had barely gained a toehold back then as compared to the present. Currently, Florida has more endangered species of plants than any other state in the South and ranks third in the nation, behind Hawaii and California. In his time, Small mourned the loss of so much tropical plant habitat as settlers turned toward commercial tourism and wholesale beach development.

Small became a leader in establishing protocols for exhibitions at NYBG, and in his lifetime he collected more than sixty thousand herbarium specimens. He also discovered the Louisiana wild iris and collected ninety species of that plant, which he first saw "growing in a swamp as the train he was on passed by. He returned using a hand-car the railroad had put at his disposal." His official NYBG biography continues: "Small distributed 6,500 packets of [wild iris] seeds and several thousand plants throughout the world. Because the swamps in which they were growing were being drained, Dr. Small is credited with saving them from extinction."

His NYBG biography also points out that much of Small's botanical writing went unpublished, for lack of funds. Botanists at the time had to pay to see their research results in print. Two philanthropists—a Florida agricultural equipment manufacturer and an aggressive New York speculator in copper mines and railroads—did fund Small's fieldwork during the peak years of his career, and the bulk of his publications are from that period.

One of Small's final projects was an engagement with the great inventor Thomas Edison, who enlisted Small to teach him about plant sampling, identification, preservation, and labeling. Edison, who had established a laboratory and home in Florida, was testing plants to find a natural substitute for the rubber plant. The project was funded by the automaker Henry Ford and the tire magnate Harvey Firestone. After testing tens of thousands of plants, Edison came to believe that a particular species of goldenrod from Kansas might carry enough latex-like sap to be converted to the production of a rubber substitute for the automobile industry, but the project ultimately fizzled when Edison died in 1931.

Small died at his home in the Bronx in 1938.

In late February 2019, gray piles of melting snow lingered around the edges of a wide driveway flanked by century-old tulip trees barely budding. Donna Campbell and I had come to the NYBG, home to the largest herbarium in the Western Hemisphere, with 7.8 million preserved specimens and still counting. While Donna took pictures of the landscape, I asked for directions to the William and Lynda Steere Herbarium, housed in the International Plant Science Center—a modern, five-story, 70,000-square-foot structure somewhat obscured behind an elegant, Beaux-Arts building that was home to the museum where Small worked. The older building is now the largest botanical library in the Americas. The greening grounds around us were an island of calm surrounded by noisy Bronx streets, clotted with traffic.

I had never been inside an herbarium and didn't quite know what to expect, but I wanted to see, if possible, how a botanist might study preserved plants in such a setting. We made our way up a long stone staircase to the ornate entrance of the grand building, with its gilded sculpture set above a stone pool out front. The vigorous figures in the golden tableau reminded me of Prometheus above the skating rink at Rockefeller Center, but in this case, the figures were sea creatures, including a mermaid and merman, frozen in motion. Inside the front door, a guard greeted us, took our names, and then dialed an upstairs office to announce our visit, which had required written permission from Charles Zimmerman, the man who now holds the position most similar to the one that Small once filled.

The garden's annual orchid show would begin in a few days, and glamorous specimens were blooming inside glass cases in the lobby looking like New York ladies in picture hats. I was nervous when our guide stepped off the elevator, but I relaxed when the assistant curator, Matthew Pace, introduced himself. "Call me Matt," he said. Casually dressed in jeans, sneakers, and a fleece top, he was a young man not much older than Small had been when he joined the staff. He had recently completed a PhD from the University of Wisconsin and told us he had done his dissertation research on a ditch orchid found in the North Carolina mountains. Soon we were sharing stories about our mutual love for that landscape. So early in my botanical investigations, I was keenly aware of my novice status and grateful when Matt treated me as if I knew what I was doing.

I'd already gotten word from the herbarium staff that Small's original specimens of the Yadkin River goldenrod, *Solidago plumosa*, were on tour and unavailable.

"They're somewhere in Canada right now," Matt explained. Apparently authorized scientists and organizations can check out specimens like they check out books.

I asked if the herbarium might have samples of Heller's blazing star, the last plant the two young botanists were looking for on the ledge at Blowing Rock. The federal government had declared Heller's plant threatened in 1987 because of the effects of acid rain and trampling by hikers, climbers, and sightseers.

"We'll look for it," Matt said.

We took an elevator up several floors and were issued special name tags before taking two flights of stairs back down to another level. We wove our way into the enormous Plant Science Center and entered a fluorescent-lit room with high ceilings and massive rows of bulky steel cabinets that took up most of the floor space. Along one wall ran a waist-high counter. This was where Matt would bring us the portfolios of specimens, page stacked on page. He explained that he would have to find the right bank of cabinets containing the family name of the plant, and then go down the row to locate the alphabetized drawers labeled by genus and then species. In the 1930s, he said, the Works Progress Administration paid people to mount samples and type out labels for a large part of the repository. Building the herbarium collection was a labor-intensive process.

If I hadn't read Small's account of what it took to secure plant specimens—the hiking, the gathering, the drying, the pressing, the storms and snakes, the sheer determination required—I might not have felt as I did in that moment, taking in the weight of all that human toil along with my own buoyant excitement in being there. The plants—collected over decades with so much effort to keep them safe—reside in a climate-controlled environment that cost millions of philanthropic dollars to build. (Government funding for this work is still scant.)

Matt went down an aisle touching each label as he walked. He finally came to the cubby he was after and opened the door. He slid out a raft of oversized papers wrapped in a dark folder, came back, and set it on the countertop. He pulled back the heavy cover. Here was a single, withered sample with the species name—*Liatris*—written with a flourish below the root end of the specimen. The handwriting was elegant and the inscription simple: "S. & H., 1891." Small and Heller. My eyes welled up suddenly, as if I were in a theater and the audience had jumped to its feet, riotously applauding the botanists bowing on stage.

The specimen was a thick, brittle, greenish-brown stalk with a feathery plume, the blossoms more or less flattened on the paper with stiff white ribbons that were glued on either side of the stems and branches to secure them in place. The New York Botanical Garden logo was rubber-stamped on the paper in black ink. Another printed label affixed to the paper read:

Plants of Western North Carolina,
Collected on the Ledges of the Blowing Rock, Caldwell Co.
August 6, 1891
By J. K. Small and A. A. Heller
Elevation 4200 feet

This date was two days before the journal entry that explained how the young botanists nearly broke their necks trying to find more specimens of Heller's namesake plant on the ledges of Blowing Rock without success. What they had found on 6 August was another species, *Liatris gramnifolia*, but that didn't matter. I felt as if we were right there on that ledge with them.

I took a deep breath. Donna took pictures. Matt smiled. So much history on a single page.

I thought of those young men and all that followed in their lives—the bright bit of fame that would forever lift Heller's name in the world of taxonomy. I imagined the spirit of Small floating around on those storied grounds where he came to work every day of his professional life. I thought of Alan Weakley, the seasoned professor with his many awards, back home at the North Carolina Herbarium. Lives all linked by plants and a shared passion for the natural world.

2

Florida Torreya

Florida is an anomaly in the South, an odd geographic appendage that is long, flat, and swampy, surrounded by saltwater, dense with unusual species, and tropical in climate. Entrepreneurs have long cast Florida as a place of secrets. The earliest would-be developers from Spain claimed they had discovered the fountain of youth here, and perhaps they did. Over time Florida's identity has been shaped by youthful images of spring-break abandon on sugary beaches and fantasy family vacations curated by Disney. So-called snowbirds from the North were early adopters of Florida as a retirement destination. They seemed to believe that the aches of aging might be reversed with year-round warm water and sunshine, advantages that state voters sweetened in 1924 with a constitutional amendment banning a personal income tax on residents.

Early in the twentieth century, as resort developers were snatching up beach properties, backwoods entrepreneurs cooked up outlandish schemes to lure tourists to stop at the rough edges of rural Florida on their way to the ocean. When I was a kid in Atlanta, my family would load up the VW bug with beach chairs, towels, suntan lotion, and a picnic of pimiento cheese sandwiches and deviled eggs. We'd hit the road for a few days of beach time in the Sunshine State. It seemed the closer we got to the state line, the more outlandish grew the billboards' promises of amazing scenes ahead. Live bears and alligators in cages, collections of venomous snakes, and fantastic replicas of prehistoric dinosaurs and sharks: all such advertised attractions were designed to seduce bored children like me to pester the driver to stop the car and take in the show.

Other destinations were more exotic: the glass-bottomed boats of Silver Springs; the world's smallest police station (a telephone booth) in the tiny town of Carrabelle, on the Gulf; and the "live mermaids" of Weeki Wachee, a natural spring where brave performers—discreetly breathing through air tubes—wore one-piece, sequined swimsuits outfitted with a single broad tail fin. My family took in a few of the old sideshows over the years, but we missed "The Garden of Eden," which opened in the early 1950s in Bristol, Florida: admission one dollar and ten cents.

Florida's Garden of Eden was the creation of Elvy Edison Callaway, a Baptist lay preacher, lawyer, and unsuccessful Republican candidate for governor in 1936. Through divine inspiration, Callaway came to believe that the "cradle of civilization" as described in the Old Testament had not been located at the convergence of the Tigris and Euphrates. Instead, he claimed, God began the human family at the confluence of Georgia's Chattahoochee and Flint Rivers, which become the Apalachicola River just south of the Georgia state line. The Apalachicola then bisects the Florida Panhandle on its way to the Gulf of Mexico.

Adding to the allure of the region, Callaway convinced visitors that a certain local tree was "gopher wood"—the material described in the Bible that Noah used to build the ark. Callaway said his story was confirmed in 1952 in the *Orlando Sentinel*, where a University of Florida forestry professor declared: "The wood in question is unique. So far as is known, it grows only on the east bank of the Apalachicola River in a small area near Bristol. There are related species to be found in China and northern California, but botanists know of no similar wood anywhere else in the world."

The presence of this unlikely evergreen growing on the steep, sandy bluffs along the Apalachicola River meant that Callaway had a ready-made nature exhibit for his tourist attraction. He erected billboards directing travelers to Eden. He fashioned a gated entry and laid out a trail through the forest, tacking interpretive signs to trees along the path. The Garden didn't last long, however. All that remains today are black-and-white publicity photos of curious visitors encountering Callaway's version of Eden on a sequence of signs that read:

BIRTHPLACE OF ADAM
HERE IS WHERE GOD DISCOVERED THAT ADAM WAS LONELY
LADIES, ON THIS NATURAL OPERATING TABLE GOD TOOK A RIB
FROM ADAM'S SIDE AND MADE MOTHER EVE
WHERE ADAM AND EVE BUILT THEIR FIRST HOUSE

And beside a shady picnic shelter that Callaway had built:

WHERE NOAH MADE THE ARK OF GOPHER WOOD

For a brief time in the 1950s, a wooded site above Florida's Apalachicola River was opened to the public and promoted as the site of the biblical Garden of Eden. The proprietor of this venture, E. E. Callaway, created a series of hand-painted signs connecting stops along a footpath to events in the book of Genesis. The signs were tacked up on native Florida Torreya or "gopher wood" trees. Photo taken in 1953 by Red Kerce, from the Florida Memory Collection in the State Library and Archives of Florida.

Of course, locals in the Florida Panhandle already knew about Calla-way's "gopher wood" tree. Some called it the Florida nutmeg because of the tapered shape of its seeds; others referred to it as stinking cedar, or "the polecat tree," because of the strong scent of its needles and seeds. For more than a century, European settlers who displaced the Indigenous inhabitants of the region harvested its lightweight, rot-resistant wood to make shingles, fence posts, and cabinets.

During the agricultural boom preceding the Civil War, they burned the wood to power steamboats ferrying bales of cotton down the Apalachicola. The fragrant evergreens also sometimes served as Christmas trees. Nineteenth-century scientists knew the tree as *Torreya taxifolia*, a prickly member of the yew family, which an amateur botanist, Hardy Bryan Croom, discovered in the Panhandle in 1833.

Croom, a wealthy planter from New Bern, North Carolina, came to

Florida in his prime to pursue his twin passions of gardening and botanizing. Croom sent cuttings and seedlings of the unusual tree to the distinguished botanist John Torrey in New York to verify that the species was previously unknown. After some debate, the tree was classified and officially named in 1838. Following Croom's wishes, both the Latin and common name of the tree honored Torrey, who would soon become New York's state botanist. Torrey's former apprentice and closest colleague — Asa Gray, of Harvard University — wrote in his journal some years later that one of the seedlings Croom sent to Torrey had been planted and kept alive for a time in New York's Central Park, "showing its aptitude for a colder climate than that of which it is native." His remark foreshadowed a controversy that has come to surround the Torreya tree's viability in Florida — to be discussed later in this chapter.

It is tragic that *Torreya taxifolia*, one of the oldest remaining tree species on earth, was among the first to land on the federal list of endangered species after having survived more than 175 million years. Today the Torreya — generally pronounced "TOR-ee-ah" by Florida locals and "Tor-RAY-uh" by outsiders — is considered the most endangered tree in North America. Sheila Dunning, of the University of Florida, writes that the species apparently migrated south into what is now Florida during the continent's last Ice Age, when "retreating icebergs pushed ground from the Northern Hemisphere, bringing the Florida Torreya and many other northern plant species with them." Six hundred thousand trees once thrived in this pocket refuge along the Apalachicola River, but now fewer than six thousand survive.

Like so many trees in the wild South, the Torreya was overharvested. By the 1940s, the population was down to some four hundred thousand. Today the tree is battling habitat loss, climate change, and a fungal pathogen that scientists began investigating as early as 1967. It took more than four decades to draw a bead on one particularly deadly fungus among a dozen that have attacked the tree over the years. University of Florida forest pathologist Jason Smith isolated *Fusarium torreya* in 2011, identifying it as an invasive blight, possibly from China, that causes canker disease.

In recent years, trees already weakened by alternating assaults of drought and ever-stronger hurricanes have been further damaged by the *Fusarium* fungus, which stunts tree growth and ultimately halts seed production. Where there were once sixty-foot-high Torreya trees in the Panhandle, roots of old trees sprout saplings infected by stem cankers and needle blight. Most new-growth Torreya trees today have topped out at about fifteen feet. Near the site of Callaway's Garden of Eden, remnant trees are protected in Torreya State Park and in Apalachicola Bluffs and Ravines Preserve, managed by the Florida Nature Conservancy. Other wild trees survive on private land in three contiguous counties in Florida and one county in Georgia.

The Torreya Keepers, a nonprofit group of local citizens, is attempting an accurate census of the remaining trees on private land in the native range: that is, the eastern slope of the Apalachicola watershed. The Keepers have pledged funds to protect the trees and to support strategies that might enlarge the native population until a blight-resistant Torreya tree can be introduced in the native habitat.

Early on a gray February morning, Donna and I set out from Tallahassee to witness an ongoing Torreya preservation project staffed on that particular workday by five professionals from the Atlanta Botanical Garden (ABG), two staff members from the Jacksonville (Florida) Zoo and Gardens, and two local landowners who are Torreya Keepers.

Forester Laurie Blackmore, the conservation and research manager for ABG, led the team. Laurie, a no-nonsense scientist with a great sense of humor, has logged years of field experience in conservation and sustainable development with nonprofits in the United States and Africa. Laurie was working closely with Jessi Allen, a bright newcomer to the ABG staff who digitized more than one hundred thousand specimens in the lichen collection of the NYBG while completing a PhD in biology from the City University of New York. In 2015, she and her PhD adviser discovered a new Appalachian lichen and named it after the country music legend Dolly Parton. Now, in addition to her present job at ABG, Jessi leads the lichen specialist group of the International Union for the Conservation of Nature.

Field biologist Ashlynn Smith drove over that morning from Deer Lake State Park, near Panama City, where she was working on a wetland habitat restoration project for ABG while finishing her PhD at the University of Florida. Two other ABG staffers—field biologist and conservation assistant Lila Uzzell and succulent specialist Trey Fletcher—were also on hand for the day of fieldwork.

When Donna and I arrived at the team's gathering place, the ABG scientists were in a huddle, talking a mile a minute with their eager Jacksonville Zoo and Gardens colleagues: horticulturalist Houston Snead and sculptor and exhibit fabricator Kyle Newsome. Meanwhile, Helen Roth, a longtime Florida Native Plant Society activist, and Anne C. Schmidt, a veteran conservation biologist—neighbors on whose adjacent properties we would work that day—were circling the core group like mother hens, making sure that everyone had packed water and snacks for the foray. Annie's rescue dog, Irish—a short-legged, wire-haired, salt-and-pepper mutt—was outfitted in a bright orange collar and matching jacket so the group could keep an eye on him as we hiked. As Laurie Blackmore would tell me later, these expeditions were always a moving botanical swap shop, with participants eagerly trading recent findings, showing off their gear, and helping each other navigate sometimes-rough terrain. We fell right in, not knowing how steep and gnarly the landscape would become.

Our mission for the morning was to plant a collection of fresh Torreya seeds taken from trees that ABG had cultivated from tree cuttings in the local Torreya preserve. As part of the multiphase effort, the cuttings were carefully grown to reproductive maturity in an undisclosed location in north Georgia, away from the native Florida habitat that harbors the fungus.

Donna and I, the interlopers, climbed aboard Annie Schmidt's trusty Isuzu Trooper, which was battered and dusty from her ongoing conservation practice. As a statewide consultant, Annie is known for her deep knowledge of Florida's native plants. She is also a certified "burn boss" for the prescribed fires that increasingly are used here to restore forest habitat.

Irish wormed his way between the front seats to nudge Annie's arm, and she threw the truck into gear. We rode on paved highway for a short distance before bouncing onto a sandy lane flanked by Helen's property on one side and Annie's on the other. A few strands of Spanish moss swung back and forth from Annie's rearview mirror as we jounced over troughs and rises in the scruffy landscape.

"Millions of years ago, this landscape was the place where sea levels were at their highest before the ocean retreated. I have three hundred acres, Helen has a hundred, and another friend, Susan, has three hundred. Together we have our own seven-hundred-acre preserve, and when sea level rises again, we'll be on the beach!" Annie laughed. Petite and wire-thin, she displayed tremendous physical energy. She went on to explain how she grew up in a large family that spent summers on Martha's Vineyard in Massachusetts. After her first marriage ended, she found and married her true love, Jack Stites, an avid outdoorsman. Early in his career, Stites was a BBC commentator on the international motorcycle racing circuit, then moved into antique boat restoration, and finally took up land management and conservation. He worked for the Florida Nature Conservancy, and for eleven years he and Annie lived on the Apalachicola Bluffs and Ravines Preserve, where ABG also has several Torreya restoration sites. Stites died two years before our visit, but Annie was still keeping a brisk pace in her conservation work to honor his memory.

She pointed out a gopher tortoise burrow as we rolled by. The endangered tortoise is a rare, prehistoric-looking species that is long-lived and native to the local region. As part of the conservation plan for the species, Florida residents must acquire permits to capture and relocate them when any land where they live is going to be developed. The gopher tortoise burrow is also known to provide shelter for more than three hundred other species. Such detailed legal protections for endangered animals on private property do not extend to endangered plants, we soon learned.

Annie slowed down a bit to climb over a steep berm. Once over the hump, the truck's front wheels dug into deep sand. She stopped, threw the vehicle into four-wheel drive, and cut a half-doughnut in reverse to make space for the others to pull in beside us. We stopped and climbed out into the damp morning. It was utterly quiet. No insects or birds seemed to be stirring. Only the dry leaves on short oaks rattled when a slight breeze came up.

The team arrived and began unloading the vehicles, and Jessi, the lichen specialist, came toward us carrying a cooler. She set it on the ground and opened the lid to reveal piles of plastic bags. Each bag was labeled with a code and contained sphagnum moss intermixed with the nutmeg-shaped Torreya seeds. "Some of these seeds are stronger than others," Jessi said, holding up a sample for Donna's camera. "Some may have turned kind of moldy and black. We want to pick three good seeds from each of the seven mothers in Georgia."

As we would learn later from Emily Coffey, ABG's vice president of conservation and research, tracking maternal-line plants for propagation is a technique that has been around for some forty years, both in plant conservation and in agriculture. Tracking seeds by the mother plant and its individual genotype "is insanely important for conserving genetic diversity," as Emily put it. She explained that conserving a species must involve separating and saving seeds that are genetically diverse across the species' genome. It is not helpful just to gather seeds from a population of plants in bulk. As a species' gene pool evolves and changes, different plants will have different tolerances and abilities to survive and adapt to variances in soil acidity, drought, and other environmental pressures. ABG drew on seven different matrilineal lines of Torreya trees and harvested thousands of seeds, Jessi said, fingering her final selections and loading them into separate bags to take into the field.

I asked Jessi if it might be possible to develop a disease-resistant ver-

Gathered in Florida on a research mission, Atlanta Botanical Garden staff members and volunteers Trey Fletcher, Annie Schmidt, Jessi Allen, Laurie Blackmore, Houston Snead, and Kyle Newsome spent the day planting, marking, and monitoring Florida Torreya tree sites.

sion of the Torreya tree using the relatively new CRISPR gene-editing technology I'd heard about on National Public Radio. "CRISPR sounds easy from the outside," Jessi said, "but we don't even have a complete genome for Torreya. Conifer genomes are enormous. They are even larger than the human genome." (A genome is the complete list of the nucleic acids and their sequence in an organism's DNA.)

Laurie, who was listening to the conversation, stepped forward. "Gene editing for the Torreya might be way down the road, but we are in agreement at ABG that we want to proceed slowly. The first question—before manipulating the species genetically—is to find and assess the genetic diversity of all the individual trees across the native geographic range right here in Florida. Around the edges of a habitat are where you tend to get the most genetic diversity, so we are searching for trees on private property at the edges of the known Torreya range to ensure we capture the full genetic diversity of the species."

Neither Annie nor Helen had located Torreya on their land, though some specimens had been found within a mile's distance on neighboring properties, making it likely that we were within the tree's historic geographic footprint. Successfully planting a diverse selection of seeds here would likely reveal more data about optimum conditions that Torreyas might need, going forward. Studies suggest that Torreyas grow slowly, need shade at first and then at least morning sunshine, and tend to be smaller than other trees in this rather odd Florida habitat, which contains both northern and southern tree species, such as beech, hickory, loblolly pine, magnolia, sugar maple, spruce pine, sweet gum, and tulip tree. Eons ago, Torreya trees were part of an ecosystem that might have included large, plant-eating mammals—grazing mastodons and mammoths, now extinct—along with ample numbers of deer and rabbits. The larger mammals kept the forest understory open, thanks to their appetite for limbs and shrubs. In past decades, the Torreya also enjoyed moderate regional rainfall and cooler temperatures at this elevation, but not so much anymore as the local climate has warmed.

To make matters worse, in 2018 Hurricane Michael—blowing at Category 5 strength some sixty miles inland—drove straight up the Apalachicola River and took out nearly three-quarters of the tree canopy in nearby Torreya State Park. Winds whipped around and snapped trees in every direction. By some estimates, Hurricane Michael took out 10 percent of the remaining wild Torreya population in a single day. Much bigger tree species fell on the native Torreya and essentially buried them, also making sensitive habitats extremely difficult for foresters to maneuver during rescue efforts.

Because Torreya trees can take twenty years to produce cones, a successful regeneration of the wild forest after Hurricane Michael has become even more urgent. With mature trees toppled, juvenile trees are less likely to make it to maturity because of the loss of overstory. They are also subject to deer grazing and antler rubbing that damages bark. Even feral pigs have been known to root out young trees. For now, treating the saplings with any kind of fungicide is considered too risky to other animals and plants in the ecosystem. In short, the Torreya tree is going away just as surely as E. E. Callaway's Garden of Eden met its demise.

Starting in 1990, ABG began developing partnerships with botanic gardens, arboretums, and universities in the region, enlisting their help to grow genetically distinct lines of Torreya trees and to gather seeds from them for preservation. In this process, they learned that Torreya seeds in the cones cannot simply be stored in dry conditions indefinitely; they won't survive. Instead, the plant embryo, which is attached to the seed, must be removed and cultured *in vitro* to be preserved. These embryos can then be kept in a cryogenic storage system for future planting. Because the Torreyas require both male and female trees to produce seeds, horticulturists must ensure that saplings of both sexes are planted for seed production. The sex of infected trees in the wild is hard to determine, though, because reproductive structures seldom show up when the tree is sick.

Despite these many challenges, a few commercial nurseries have been selling Torreya trees online, and some vendors assert that their trees are disease-free. Laurie was doubtful. "The *Fusarium* fungus takes a while before it manifests—two or three years," she said. "There's still so much we don't know, and we have limited funding to do this research. The only good thing coming out of Hurricane Michael was the federal funds that have been made available to us to advance this study. Working with private landowners like Annie and Helen is the heart of our project, and we are now hiring a local person to work with property owners to identify more Torreya trees on private land around here."

With all the gear in hand, the crew followed Annie down a gently sloping trail through pines that had been thinned some years back. She explained that her property had belonged to the St. Joe Paper Company and was cleared for timber. "St. Joe never owned Helen's property, so it was left untouched," Annie said.

As we got deeper into the forest, we could see the hurricane devastation—trunks of pines and gigantic oaks twisted and felled by wind. We approached a narrow, man-made bridge and crossed single file over a crystal-clear stream with a sandy bottom. I followed Irish onto the bridge, but he soon jumped into the water and began winding his way up the creek bed in his orange costume, now wet around the edges. "This is Crooked Creek," Helen said.

Beyond the shimmering creek there was no obvious trail. Annie began climbing the steep slope like a billy goat and wove through the tangle of vines and shrubs that soon grabbed at all of us. Keeping up with her was hard. We wound across the slope, up and down, according to breaks in the winter foliage just wide enough for us to get through.

I was grateful not to be carrying the gear bags or the cage made of hardware wire that would be used to cover the new seedbed. Trey, the succulent expert, and the guys from the Jacksonville Zoo took turns shouldering the equipment through the snags and snares along the slope. We tried to hold back the whiplike vines and branches, shepherding each other through ragged tunnels of shrubs without smacking each other in the face by a limb let go too soon. Annie, in the lead, just kept moving ahead, and Irish, now both wet and sandy, ran ever wider circles around us.

I am glad that on that hike I had not yet discovered this passage in the book *Florida's Uplands*: "One naturalist has observed that black widow spiders in the Apalachicola ravines live in trees and shrubs, whereas elsewhere in Florida, as far as we know, they live on the ground." More encouragingly, the authors went on to say: "The single 35 mile stretch of ravines on the east side of the Apalachicola River in Florida harbors more total plant and animal species, and more endemic species in particular, than any other area of the same size on the southeastern Coastal Plain. The ravines are home to more than 100 rare and endangered species. Despite their fame, however, the Apalachicola Bluffs and Ravines have never been exhaustively surveyed; they have many more secrets to reveal in times to come."

I can see why a thorough exploration has not yet been managed, and it also registered that we were in sacred space—a truly novel place on the planet. Helen Roth, who dropped back to join us at the rear of the group, told us how she became active in plant conservation in 2009, when she and her husband, Tom, bought this tract from Helen's father and named it Spring Canyon. Laurie kidded Helen, telling us that her neighbors have started calling her Chain-Saw Helen for her brush-clearing talents. In 2012, Helen signed up to be a part of the Working Lands for Wildlife Program, which restores habitat for gopher tortoises. With a few helpers over the years, Helen has cleared some thirty-eight acres and is also certified to manage prescribed burns as needed on her property.

The Baptist church that owned the land before Helen's family acquired it had dammed Crooked Creek to create a pond, and native hardwoods that are not resistant to fire were able to encroach into the fire-dependent longleaf pine and wiregrass savannah. Helen's clearing and carefully timed burnings have now created room not only for a resurgence of the gopher tortoises and fox squirrels but also for the natural regeneration of old-growth longleaf pines. Native flame azalea, mountain laurel, and magnificent magnolias (*Magnolia grandiflora*) thrive within the cooler temperatures of the steephead ravines. Such plants, best known for their residence in the Appalachian Mountains, perhaps explain how Apalachicola got its similar-sounding name. We stopped for a moment to take in the view around us. Judging by the surrounding trees and shrubs, it could have been a wild spot just off the Blue Ridge Parkway in North Carolina, except for the February mosquitoes, the humidity, and a temperature near 80 degrees.

Helen told us proudly that the management of her piece of the forest has increased the population of the endangered Gholson's blazing star, also called Bluffs gayfeather (*Liatris gholsonii*). This rare species of liatris, which is kin to Heller's blazing star discussed in the previous chapter, grows well here. Florida State University professor Loran Anderson discovered the species in 2001 and named it for Angus Gholson Jr.

Gholson was a revered citizen-scientist whose reputation for botanical knowledge and great storytelling brought visitors from all over the world to his house in nearby Chattahoochee, Florida, where he was born, raised, and lived with his wife during their seventy years of marriage. Known to be especially fond of simple country-style cooking, Coca-Cola, and most any dessert, Gholson once told a reporter for the *Sun Sentinel* newspaper: "Three things are very important: air, water and food. You can't do without air but for just a damn few seconds. You can't do without water or food for long, either. All of it comes from plants. I don't know what we're going to do when we cut them all down. We're going to have to plow up some of these interstates and plant corn." Gholson died in 2014, at the age of ninety-two.

Helen Roth intentionally adopted Gholson's model, learning and sharing stories of her surroundings, giving forest stewardship tours, and welcoming native plant enthusiasts and other landowners to her upland pine forest and steephead ravine. According to a recent survey of the Torreya seeds that she has planted in a variety of locations at Spring Canyon, she has eighteen seedlings so far that have germinated and are growing.

As our group straggled along the slope, Annie called out to Laurie to check our bearings. Laurie pulled out her phone and punched up a global positioning app called Avenza that allows users to choose any kind of digital map—topographic, geographic, geologic—to track their path in a terrain. Users can drop a pin to label sites that they want to record and remember, all without benefit of an internet connection. Laurie could see on her phone that we were close to one of Annie's experimental Torreya beds created by the team several years back. She pointed, and Annie headed straight to it. A tree had fallen on the first set of caged Torreya seedlings, but they had survived. Annie was soon kneeling in the duff, removing leaves and sticks to uncover a second planting, put in without benefit of a cage for protection. She hooted when she found several more seedlings.

This slow-growing Florida Torreya seedling (Torreya taxifolia) was planted by the Atlanta Botanical Garden on private land near Quincy, Florida. It managed to survive the devastation of Hurricane Michael.

"It's amazing!" she hollered, as the rest of us caught up and saw the two- and three-inch seedlings rising from the salt-and-pepper soil.

The team got to work. Trey's task was to clear leaves to mark a third bed, four feet by three feet. Each Torreya seed from the seven mothers would get a small hole dug with a shiny dibble marked to a precise depth, in hopes that one or more would survive. A metal ID tag would be pinned to the soil beside the buried seed. Meanwhile, Jessi had started clipping an opening in the center of the wire cage where she would attach two devices: a soil probe that measures moisture and soil chemistry and a self-contained data cell that measures sunlight and temperature every thirty minutes and relays the information by Bluetooth to her phone. She handed me the data cell. It fit in the palm of my hand.

"The battery in this will last 2.7 years if it logs the information on the half-hour," Jessi said. "If I set it up to measure every ninety minutes, the battery will last eight years. We'll come back in the fall to check on all these instruments."

Nearby, Laurie was balancing in her outstretched hand what she explained was an old-school forester's tool. The "spherical crown densitometer" had a small, rounded mirror that reflected the sky overhead and measured the dome of light that was coming through the tree canopy. From her device, Jessi reported that the ambient temperature was 79 degrees and the light coming through the tree canopy was 7,400 lumens. Yet another instrument measured photosynthetically active radiation, which is necessary for plant growth. Jessi said all of these data points would be collected as the seeds germinated and would help the team understand what is important for Torreya growth. They hoped to learn which of the seven maternal lines might produce the most robust seeds as they adapted to different microclimates among the seedbeds scattered throughout the forest.

Several years back, Emily Coffey had picked the nine sites for this study: five locations on public land and two each on Annie's and Helen's land. "Seed trials on private land can be challenging," Laurie explained. Government regulations and policies about endangered species can facilitate conducting experiments in public spaces. At the same time, rare plant species growing on private land belong to the landowners. "They can do what they want to their flora, even in the case of endangered species, and they can plant invasives at will," Laurie said. Finding study participants like Annie and Helen to permit this work on their property was a great gift. They accepted the risk that the *Fusarium* fungus might be present in the seedlings already growing there or in the seeds being planted that day.

At ABG, a row of Torreya trees planted more than twenty years ago flanks the edge of the tony Piedmont Driving Club, next door. Those trees are descendants of cuttings that Harvard University's Arnold Arboretum shared with Atlanta in 1989. All have been infected with the *Fusarium* fungus even as they are cared for fastidiously by ABG staff, suggesting that the fungus is likely more than a lone invader; its success may depend on conditions in the soil, the surrounding ecosystem, and the relative hardiness of the individual tree when it is attacked.

Botanists constantly worry that the *Fusarium* fungus could violate other tree species or plants in new habitats. Regional conservationists in particular wince at the memory of the American chestnut tree (*Castanea dentata*), which was infected by a blight accidentally imported to the United States from Japan in the 1890s. Identified in 1904, that fungal pathogen spread through eastern hardwood forests and killed billions of trees over a fifty-year period. The beleaguered chestnut, so beloved and attached to the identity of the Appalachian region, inspired the Southern Appalachian Botanical Society in 1936 to name its scientific journal *Castanea*.

Botanists who have witnessed the dwindling Torreya forest in Florida have noted that the spindly appearance of infected saplings is similar to the American chestnut stumps that linger in southern forests. To this day, the blighted chestnut remnants will send up shoots, but they never mature to the height or glory of their forebears.

Time will tell for the new Torreya seeds now planted in their native habitat. The dedicated conservation work of ABG and its many cultivation partners, however, is not the only work that is being done on behalf of *Torreya taxifolia*.

Torreya trees, it turns out, thrive in some surprising locations far north of their native Florida habitat. Torreya trees have been planted in years past for the novelty of the species and more recently in direct response to the tree's decline in the wild. One of the Torreya's most dedicated advocates is Connie Barlow, who divides her time between Michigan and Florida. She describes herself as an "evolutionary humanist." In the 1990s, Barlow was a writer of popular science articles, first as a contributor to *Wild Earth* magazine and later as the author of several books. Two of Barlow's titles were published by MIT Press and focused on the idea of a more harmonious relationship between Christianity and the science of evolution. With her husband, Michael Dowd, an eco-theologian and former pastor, Barlow has promoted a "deep time" understanding of the origin story while advocating a brand of children's religious education that emphasizes stewardship of the earth. Once called "America's evolutionary evangelists," Barlow and Dowd created a podcast by the same title and have traveled the country promoting humanity's sacred obligation to future generations by protecting and sustaining nature.

In particular, Barlow has devoted many years to raising awareness of the plight of the Florida Torreya tree. In 2004, she joined with the late Paul S. Martin (a University of Arizona Pleistocene ecologist) to establish Torreya Guardians—not to be confused with Annie and Helen's group, the Torreya Keepers. The Guardians are also a volunteer association of citizen naturalists with a mission to "rewild" the Torreya tree in habitats outside the native range—a practice that they believe will save the species.

The Guardians began by taking inventory of surviving *Torreya taxifolia* that were outplanted beyond their native range in Florida—from the North Carolina mountains to points much farther north and west. For their initial rewilding project, they acquired 110 seeds from a grove of Torreyas that was planted in 1939 on the elegant Biltmore Estate in Asheville, North Carolina. Barlow and her team are convinced that the tree has already proven its capability to survive in the Appalachian chain, where, they hypothesize, the tree might have originated before glaciation. In 2008, they planted thirty-one Torreya seedlings grown in a South Carolina nursery on private land near Waynesville, North Carolina, in the Blue Ridge.

The Guardians' website contains Barlow's meticulous and ongoing documentation (both narrative and video) of the group's plantings and observations. She and others travel from site to site, monitoring tree growth and setbacks. To date, the Guardians have also made seeds available to educational institutions and botanical and conservation organizations across nine states and have recruited planters on private properties in twelve states, including sites as far north as Ohio, Michigan, and New Hampshire. The organization does not have bylaws or a formal board of directors. Rather, each activist is encouraged to work independently, following the ecological standards and horticultural practices for planting cuttings and seedlings that Barlow and her colleagues have developed.

The term "rewilding," initially used by the Guardians to describe their early plantings of Torreya outside of Florida, was a disturbing term for many mainstream botanists. "Assisted migration" is now the preferred phrase in scientific literature.

Vivian Negrón-Ortiz is the Panama City–based Fish and Wildlife Service (FWS) botanist who is charged with monitoring the tree's endangered status and fulfilling the conservation plan developed for the US government. She has been reluctant to endorse any remediation outside the native habitat. As she explained in a document produced by the southeastern regional office of FWS: "We have to look at all alternatives to have the species *in situ* [in its original habitat] conserved and protected. Assisted migration could be an alternative given climate change and if there are no other options. But we have to have a plan in place first. It will probably take a lot of resources and a suite of partners to save the Torreya."

ABG's Emily Coffey agrees. "Assisted migration ought to be the last resort," she told me when we met, noting that "the Torreya does not have a place in the Appalachian ecosystem, so introducing these trees without rigorous scientific testing would be irresponsible. I am not opposed to assisted migration if it is able to show with data that we will not be introducing pathogens into another ecosystem." Some of the long-lived trees in Biltmore Forest have shown signs of the fungus.

For her part, Connie Barlow is frustrated by the snail's pace of formal scientific study alongside the bureaucratic procedures of governmental officials maintaining compliance with the Endangered Species Act. Barlow argues that we are in a period in which certain plants have become ecologically out of step with changing climates and habitats and must be rescued by human intervention, lest they become extinct. But as a conservationist, Barlow is equally rigorous in acknowledging concerns about assisted migration, even though there are no laws against it. In an essay collection called *Gaia in Turmoil: Climate Change, Biodepletion, and Earth Ethics in an Age of Crisis*, Barlow wrote: "Assisted migration frightens for precisely the same reasons it fascinates: anybody can do it, for good or ill, and with care or abandon. Its promotion could undermine decades of public education about the dangers of nonnative plants, as well as more recent efforts to promote the concept of wildlands corridors and connectivity."

Science writer Douglas Fox interviewed scientists about the specific and varied dangers of assisted migration for the journal *Conservation*. Some of his subjects worried that a species carried away from its native habitat could become invasive in the new one. Stephen Rice, of Union College in Schenectady, New York, explained how black locust trees that voluntarily migrated north into New York and Wisconsin from their native habitat in the Appalachians soon began edging out some native plants in the new habitat. Over the long term, the presence of these locust trees will likely enrich the soil, Rice added, which will in turn invite more invasive plants to move in and displace native plants that had adapted themselves to the poor soil.

At the other extreme is the possibility that trees moved with human assistance will simply die in their new environments. The migrant trees could also interact with other species in the new habitat and create hybrids that will outpace and even kill the original tree that the planters were trying to save. Because of climate change, Fox concludes, it is certain that tens of thousands of species "will move whether or not humans do the moving. And many currently protected habitats will become populated with exotic species. The question is whether newly populated habitats will function at least on a basic level."

It must be said that hybrids occur in nature all the time, often to positive results—creating a better adapted species that can play a beneficial role. "Plants have always moved where there is opportunity," said Carol Reese, an extension horticulture specialist housed at the University of Tennessee's West Tennessee AgResearch and Education Center in Jackson. Carol is also an occasional contributor to the lively website GardenRant. "The only constant on the planet is change. Plants have always moved in where there is opportunity, and that change creates other opportunities for change, which is evolution in action. Adaptation. Story of the planet. Period," she said.

The hybridization of Torreya in the eastern United States is unlikely, because its nearest relatives live on the West Coast, and there appears to be a relatively low risk of Torreya becoming invasive in a new ecosystem. The real danger would be from any pathogens it carries with it, such as the *Fusarium*.

In 2018, ABG and the University of Florida hosted a two-day meeting called "The Tree of Life Conference." In addition to talks and panels on the Torreya, participants took a trip to the Apalachicola Bluffs Preserve for a ceremonial planting of a Torreya seedling by the eminent biologist E. O. Wilson, then in his late eighties. Wilson told the group that he had first visited the Apalachicola Bluffs in 1957, expressly to see the Torreya, "the way you would go to Paris to visit a cathedral." Wilson then declared the site to be "not only a cathedral, but also a battleground where one of the greatest events in American history will take place," meaning the battle to preserve the earth's biodiversity. That was in March 2018. Seven months later, Hurricane Michael brought its army of wind and rain, burying Wilson's ceremonial seedling.

Because of the controversy around assisted migration, the partner institutions that have been cultivating *Torreya taxifolia* for ABG have operated with a serious degree of secrecy. Before traveling to Florida to witness the planting of ABG seeds, I met with Annabel Renwick, curator of the Blomquist Garden of Native Plants at Sarah P. Duke Gardens, part of Duke University in Durham, North Carolina. Renwick oversees a hidden orchard of Torreya seedlings growing as part of the ABG cultivation project in the university's Duke Forest—a seven-thousand-acre mixed pine and hardwood forest near the campus that is used for research and teaching.

To protect the Duke orchard from vandals or other disturbances, only Duke staff know the location. Like a CIA document, descriptions of the nursery and orchard site are blacked out in copies of a master's thesis that provided the site plan for Duke's project.

Annabel introduced me to Katherine Hale, who conducted the planning and research for the orchard as part of her work toward a master's degree from the Field Naturalist program at the University of Vermont. In collaboration with Duke Gardens, Katherine traveled a thousand miles around the South to investigate Torreya trees, including those cultivated by ABG. She visited the older plantings of Torreya in North Carolina and private properties where Connie Barlow's Torreya Guardians planted seedlings in the past decade. She documented other singular sites of Torreya in South Carolina, Georgia, Tennessee, and eastern North Carolina. Her goal was to determine the conditions that would make for strong seed production in Duke Forest.

Katherine is from Durham, where her mother teaches pathology and immunology at the Duke University School of Medicine. "I grew up running around Duke campus and in the woods. I was probably one of the last free-range children," she told me when we sat together in the library at Duke Gardens.

Katherine earned her undergraduate degree at St. John's College in Annapolis, Maryland, also known as "the Great Books School," where there are no tests or lectures, "just reading as much as you can, starting with the Greeks and going up to about the 1950s," she explained. She read Euclid for math class, Ptolemy for physics, and Darwin and Aristotle for biology and philosophy. I wondered if this curriculum might have been similar to what earlier botanists, such as Hardy Croom, John Torrey, John K. Small, and Arthur Heller, read as undergraduates.

"It really brought home to me," Katherine said, "how much knowledge changes: that what we take as fact is frequently overturned in a hundred years or less. Science, like the world, like the plants themselves, is constantly changing. It feels especially true in the unfolding Torreya story."

For her master's thesis, Katherine knew she wanted to grow something. Home for a summer, she volunteered at Duke Gardens and met then-curator Stefan Bloodworth and his successor, Annabel Renwick. They dis-

by our heels digging in to maintain purchase on the sheer slope—rose up rich and earthy-smelling. A pileated woodpecker called in the distance, and Laurie, Lila, Trey, Donna, and I decided at once to sit down. The pathless climb down looked too steep for my wobbly legs, already weary from the three miles up, down, and across that we had hiked with Annie, according to the pedometer on my phone. At this height, we were looking into the top of an enormous American holly tree (*Ilex opaca*) that Trey estimated was at least at sixty feet tall. Where its trunk met the soil below, we could barely see through the understory. We couldn't make out the Torreya so far down in the thicket either, but Laurie assured us it was there. We sat in silence for a time.

Donna asked, "Why do you all do this, with the odds stacked so high against the Torreya?"

"Yep," Laurie said. "That's a good question. The Torreya has been on its way to extinction for a long time, but now it's going away faster than it should. We have to do what we can." She paused. A breeze stirred, and dappled light played across the ravine.

"We have to *try*," Lila finally said.

I realized we had gone all day without seeing a Torreya tree any bigger than a seedling. Donna and I turned down the highway toward the public entrance to Torreya State Park. We were very near the site of E. E. Callaway's Garden of Eden. I noted on the map that Garden of Eden Road still runs off Highway 12 just fifteen minutes from where we were headed.

When we arrived, the state park was open but empty of visitors. The Gregory House, a two-story antebellum mansion that the Civilian Conservation Corps had moved from across the Apalachicola to this dramatic site overlooking a wide bend in the river, was closed to visitors on this weekday. The yard had been meticulously groomed. Just beyond the parking lot, as we approached the house, a state historical marker bearing the story of the ancient tree stood beside six Torreyas in a cluster, spaced with only a few feet between them. None was taller than fifteen feet; one was barely a foot tall. The cankers and needle wilt were unmistakable. I picked a fresh needle and broke it in half to smell the evergreen. It had an energizing citrus scent.

These trees were a mighty poor representation of the gopher wood that Noah must have used to build his ark while the rains threatened. It came to me then that the story of Noah shepherding all the endangered animals, two by two, aboard the boat was the first imagining of assisted migration in human history.

3

Alabama Canebrake Pitcher Plant and Green Pitcher Plant

 The North American pitcher plant is an adaptive genius and one of the most oddly alluring of all flora. In only 3 million years—a very short time in plant evolution—these charismatic perennials have adapted to poor, acidic soils by becoming carnivorous. Dressed in garish colors with dramatic, trumpet-shaped leaves laced with contrasting veins, pitcher plants entice unsuspecting insects to serve as their supper. The hood that hovers over each trumpet secretes an irresistible nectar, seducing crawling and flying insects to explore the lip of the pitcher and the underside of the hood. Visiting ants and other insects soon find themselves trapped. The upper throat of the trumpet is wickedly slick. If a visitor slips and grabs at the walls deeper inside, the stiff, downward-sloping hairs make it nearly impossible to crawl back out. Eventually the weary insect falls into a well of toxins at the bottom of the pitcher and dissolves into a happy meal of nitrogen and phosphorus for the plant. Such extraordinary attributes have also made pitcher plants irresistible to human collectors and poachers.

Alabama's two carnivorous species on the federal endangered list are the Alabama canebrake pitcher plant (*Sarracenia alabamensis*), of which there are now fewer than a dozen known wild populations across two Alabama counties, and the green pitcher plant or mountain trumpet pitcher plant (*Sarracenia oreophila*), which occurs only in northeastern Alabama and a few counties in northeastern Georgia and southwestern North Carolina.

Among the twenty-nine carnivorous plant species native to the South, botanists believe that fewer than a dozen distinct species of pitcher plants are left in the wild. What's worse, less than 3 percent of their native habitat remains. No species of *Sarracenia* has gone extinct yet, but *S. alabamensis* may be the first in line.

Various pitcher plant species have devised clever ways to avoid consuming the pollinators they need to help with their reproduction. According to Dr. Jess Stephens, the former conservation coordinator at ABG, some plants sprout flowers to attract pollinators first and then develop pitchers later to gather food. Others—including the Alabama canebrake pitcher plant—grow both pitchers and blossoms at the same time, but the luscious blooms open on tall stems far above the hood of the pitcher, distracting bumblebees and other needed pollinators away from the deadly well. "Another mechanism thought to prevent bumblebee pollinators from being victims of the plant is the chemical cues given off from the pitcher and the flower," Jess explained. The hypothesis is that the smell that attracts prey to the pitcher is different from the smell of the flower that attracts pollinators.

As if all this natural adaptation weren't extraordinary enough, pitcher plants also manage to host some seventeen other known organisms—certain mites, midges, mosquitoes, and various bacteria—that depend on the pitchers to thrive in community.

Stephens is among a recent crop of scientists who have studied these pitcher plant partners, including a species of moth, *Exyra semicrocea*, which adopts the pitcher plant as home, leaving the pitcher only for short mating flights during its lifetime. Female pitcher plant moths lay a single egg on the inside wall of a pitcher. When the larva (caterpillar) emerges, it acts like a spelunker with its own cable and harness to avoid falling into the well. It munches on the plant's inside wall, while fastened securely to a silken thread attached above the feeding site. Should the caterpillar lose its footing, it will curl and drop, suspended just short of the toxic juices below.

The caterpillar cocoons in the pitcher over winter as the plant dies back. In spring, an adult moth emerges, wearing something like moon shoes (called pretarsal claws) that allow it, too, to cling to the walls of the new-growth pitcher before taking flight. As the weather warms, male and female moths fly around outside the pitchers at dusk. After they mate, the females deposit their eggs and the life cycle begins again. Because each individual plant has multiple pitchers, it can weather the caterpillars that feed on it. Similarly, a singular species of mosquito (*Wyeomyia smithii*) has developed a dependence on water droplets inside certain pitcher plants to deposit their larvae as part of its reproductive process.

So pitcher plants are cozy bassinets for some insects and a shop of horrors for others. The bad news is that, in a recent survey of several historic southern bogs where pitcher plant moths have been observed regularly by botanists since the 1920s, Stephens and an associate documented four sites in northern Alabama that had no *Exyra* moths where they had been present in the late 1990s. In an interview with botanist Matt Candeias on the podcast *In Defense of Plants*, Stephens suggested that the pitcher plant moth may well be the canary in the coal mine. The miniature ecosystem created by the Alabama canebrake pitcher plant, like the plant itself, is in deep trouble.

Human curiosity about carnivorous plants has been perpetuated over generations. Charles Darwin wrote a definitive book on them in 1875. Even earlier, in the eighteenth century, William Bartram first encountered pitcher plants on a tramp through the South with his father. In the introduction to his classic book, *Bartram's Travels*, he wrote: "Shall we analyze these beautiful plants since they seem cheerfully to invite us? How greatly the flowers of the yellow *Sarracenia* represent a silken canopy? The yellow, pendant petals are the curtains, and the hollow leaves are not unlike the cornucopia or Amalthea's horn. What a quantity of water a leaf is capable of containing: about a pint! Taste of it, how cool and animating—limpid as the morning dew."

Just be sure to pick the mosquito larvae out of your teeth after you swallow that sip of water!

Not only have carnivorous plants unique to the wild South fallen victim to habitat loss from agriculture and human encroachment; they have also become a crime scene waiting to happen. Given that pitcher plants' remaining habitat is often in remote and economically depressed areas with poor soils unsuited to crops, human poachers are sorely tempted to gather up these expressions of nature's bizarre and beautiful creativity. Then they sell them to dealers, who offer the plants online, often at high prices.

The black market for carnivorous plant species is global. Visit eBay on any given day and you'll likely find *Sarracenia* natives of the US South offered by dealers in the United Kingdom, Italy, France, the Czech Republic, and beyond. Only one nursery is approved by the US Fish and Wildlife Service: Meadowview Biological Research Station, in Woodford, Virginia, which is permitted to sell endangered *Sarrecenia* across state lines.

Barry Rice, a professor of astronomy in California, is a highly knowledgeable and generous carnivorous plant grower, collector, and photographer. On his website, Sarracenia.com, Rice reminds readers that global regulations require sellers to obtain permits when shipping pitcher plants internationally. "It is ethically a big no-no if the plants are pulled from the wild in a non-sustainable or illegal way (as most field collection is!). But that is not the legal risk," Barry explained to me by email. Federally endangered pitcher plants cannot be sold across state lines without government-issued permits. He added, "Hobbyists who order plants by mail risk prosecution if the plants they buy do not have the appropriate permits or legal paperwork."

Pitcher plants are grown and sold legally in the horticulture trade—we saw them for sale at the genteel Birmingham Botanical Garden in our travels—but populations cultivated outside the historic range of the species are not considered natural or native, making the true native plants that are still growing in the wild the ones most sought after by some collectors. Selling plants cultivated by nurseries, we learned, helps take some of the pressure off plants in the wild, but not enough.

I was amazed to learn how many enthusiasts are out there. A longtime friend who lives less than a mile from me in North Carolina revealed that two doors down from her house is a flourishing bog established several years ago by a newcomer to the neighborhood. "He has dazzling pitcher plants and Venus flytraps that he babies incessantly," she said.

Another old friend told me that her teenage grandson had surprised her by asking for a grow light for Hanukkah this year. I raised my eyebrows and grinned, remembering the 1970s, when perfectly respectable young academics were growing marijuana in their home basements with such lights.

"Of all things," she explained, "he's taken an interest in pitcher plants."

The North Carolina Botanical Garden is not missing out on this trend either. They throw an annual party in early summer called "Pitchers and Pitchers." The staff serves up local craft beers to accompany a public program about the garden's carnivorous plant collection. Beyond the patio where the beverages are chilling, the garden's glorious paisley pitcher plants can be seen showing off in a leafy bog.

Noah Yawn is a native of Birmingham and will graduate from Auburn University in 2021. When he was ten, he told me, his grandmother gave him a Venus flytrap—probably the best-known carnivorous plant in this country. Noah was instantly hooked. "I fed the flytrap hamburger meat, and it died," he said, shaking his head. "But then I found out about pitcher plants and how they were growing in the wild about an hour away from where I lived. The first bog I visited was Splinter Hill, in Atmore, Alabama, when I was fifteen."

Noah began growing carnivorous specimens that he acquired from a local nursery. The specimens he bought were commonly available varieties that originated from tissue culture, which is an ethical/legitimate source. Noah set up a bog in a backyard baby pool at his family home. By his freshman year in high school, he had gotten a job at the same nursery that sold him the plants. He had developed such a keen interest in pitcher plants that he then got in touch with Keith Tassin, currently the director for terrestrial conservation at The Nature Conservancy of Alabama.

"I didn't realize how senior he was in the organization," Noah admitted. He asked Tassin for permission to visit a restricted area in the wild to photograph the Alabama canebrake pitcher plant. After they talked for a while, Tassin agreed and sent along a map. Noah's dad drove him to the preserve.

A couple of months later, Noah called Scott Wiggers, a senior botanist at the US Fish and Wildlife Service in Mississippi. This time the conversation was even longer. Noah wanted to know how to get involved in conservation, what he might study in college, and where to find more information. "I started reading a lot of papers and literature on *Sarracenia* and other threatened species to educate myself during my senior year," Noah explained.

With his wide smile and humble manners, Noah soon earned the trust of conservation professionals and several private landowners who had pitcher plants on their property. He launched a project on his own, gaining permission to document five natural sites of canebrake pitchers in two Alabama counties. "I noticed the odd throat patches and wanted to learn more, so I ended up doing the visiting and 'data collection' in my free time during my senior year of high school," Noah explained. Two sites were on Nature Conservancy land, and the other three were on privately owned land.

He counted the pitchers across the five sites, measured their heights, and focused on variations in external pigmentation (a fluorescent green-yellow) and pigmentation inside the throats, which have dark maroon veins. Noah published his preliminary findings in the *Carnivorous Plant Newsletter* in 2018. He also landed an offer to study at Cornell University, but opted instead for a scholarship to attend Auburn University, much closer to home.

"The article I wrote got me on the radar of the pitcher plant experts at

ABG," Noah said. "So that winter, as a college freshman, I went over and met with Jess Stephens and [pitcher plant guru] Ron Determann at ABG. I learned that the big problem with a lot of these rare *Sarracenia* is that resources are very limited for keeping deep data sets on them, and collecting data is very time-intensive. The Nature Conservancy does a phenomenal job managing their preserves, but their capacity is spread thin," he said. "The last extremely detailed full census on pitcher plants was collected in Alabama in 1995." Noah Yawn was born in 1999.

When he was invited to participate in a full census in the field, Noah was thrilled. He started the project as a summer internship with ABG. He and Jess visited all the known pitcher plant sites in his native state. Using the 1995 survey method along with a few other parameters, they wanted to see how much the populations had changed.

"Most have changed very drastically," Noah said. "Having guaranteed funding for the management of these populations is hard when they are mostly on private lands. A lot of populations have declined. We are seeing this polarization in the landscapes. You have a bog and then, right where the bog ends, you have a dense forest—usually old pastureland that has succeeded into thick forest or has been transformed into a pine plantation. There's no room for the bog to spread anymore, and it is locked into the landscape. The only way to fix that is to open it up and get fire in there." Prescribed burns can reinvigorate not only pitcher plants but also the entire forest.

Wanting to see pitcher plants in the wild, Donna and I set out from Birmingham on an early morning in May 2019, with plenty of time to get lost on the way. We planned to rendezvous with Chuck Byrd, a forester with The Nature Conservancy in Alabama. We drove west, then south, through densely forested stretches of countryside that were interrupted by recreational lakes. At that early hour of morning, the glossy waters were still, framed in a luminous fog that drifted above the surface like sheer scarves. From time to time, a V of ducks or a single heron would pass overhead. Signs directed travelers down narrow side roads to private marinas and public launches for fishing boats. Other handmade notices on telephone poles promoted upcoming catfish dinners and church yard sales.

In the first two decades of the twentieth century, northern Alabama's exceptional number of man-made reservoirs were created to produce hydroelectric power and jobs. These new bodies of water were soon embraced by sport fishermen, boaters, and water-skiers. Well-heeled city dwellers from Birmingham and Montgomery still head out of town on

Fridays to spend weekends at lakefront second homes or in houseboats, where steady breezes and long views of glassy green water are most welcome, especially come summertime in Alabama.

Occasionally we passed fields cleared for cattle and horse farms, all well-kept and expansive. The closer we came to our destination, however, the fewer gas stations we saw and not a single chain restaurant. The tiny communities we breezed through were sorely challenged, as evidenced by the scarcity of commerce or industry. For the most part, our fellow travelers on the small highways that weekday morning were heavy-laden tractor-trailer trucks hauling felled and debarked trees, all pine.

As Donna pointed out, the thick woods around us did not look like early May back home. The greenery was so dense that it looked more like July in North Carolina. But we were in the heart of tree-farming territory— mostly loblolly pine—grown for paper and other pulp products. Alabama easily ranks among the top five states in the nation in the forest industry. Pines here are planted in rows so close that nothing can grow in the dark understory—a first clue to the increasing rarity of the pitcher plants that we aimed to visit. This species once thrived in nearly full sun in the less-dense pine barrens that covered the region before it was given over to paper- and power-making. Even though Alabama's state tree is the longleaf pine, it was nearly logged out of existence and replaced with loblolly. Fortunately, longleaf is the target of new conservation efforts.

So while Alabama has a largely plant-based economy, resources to protect its unusual plant specimens are lacking. Poaching of the federally endangered Alabama canebrake pitcher plant and the green pitcher plant is rampant, according to Chuck Byrd, whom we were aiming to meet at a secret destination in the wild.

We had picked a time when the canebrake pitcher plants would be blooming. I had only seen pictures of the stunning color combination of red blooms and yellow pitchers. Once Byrd found an open spot on his calendar, he emailed us the longitude and latitude coordinates for a protected preserve in deep woods about an hour south of Birmingham.

"That's it? Just compass numbers?" I asked Donna.

"Yes," she said.

The trip felt like an undercover mission, and it made me a little nervous. Before we left home, I had plugged Chuck's compass points into Google Maps, noting first the very small rural roads that would get us there. Then I toggled over to the satellite view—a very dark, solid green blotch on the map. The destination seemed so remote that I wondered if we could actually find Chuck for the meetup. Donna realized we'd better have the turn-by-turn directions printed out on paper, in case we had no cell service as we drew closer to the dot on the map.

We drove on, seeking to understand the intense allure of pitcher plants to budding conservationists like Noah and to the unscrupulous entrepreneurs who dig them up in hopes of making a profit. According to the nature newsletter *Mongabay*, environmental crimes have become "the fourth largest criminal enterprise in the world following drug smuggling, counterfeiting, and human trafficking."

Yury Fedotov is the executive director of the United Nations Office of Drugs and Crime, which runs the Global Programme on Wildlife and Forest Crime. He told *Mongabay*, "Countries must be supported in developing sustainable livelihoods for communities and to better protect their natural heritage." Stopping the commodification of rare plants, he said, "must include strengthening customs security at ports and border crossings and increasing the use of wildlife forensic science to ensure the proper identification of species." The profits of poaching—both animals and plants—are mind-boggling, and corrupt governments in many countries support the industry, the newsletter said. It's a short-term vision, given that human life ultimately depends on these complex ecosystems.

The Alabama canebrake pitcher plant was discovered and named in 1974 by a husband-and-wife team of Michigan botanists visiting in Alabama. Some of the plants they found were growing on land owned by paper companies. Once discovered, the species was listed by the US Fish and Wildlife Service as endangered, and now it is also listed internationally in the Integrated Taxonomic Information System (ITIS), the source for authoritative taxonomic information on plants, animals, fungi, and microbes of North America and the world.

As a policy, neither of these agencies reveal the specific locations of the plants. Yet, as Noah Yawn would explain to me later, the moment the official endangered listings were published for the canebrake pitcher plant in Alabama, collectors wiped out the entire population of pitchers in Elmore County. Now only two counties have populations. "With this being published as a new species for the public, collectors could read the species paper and figure out the type locality. That is what led to the start of poaching for the species, and why Elmore County's populations were wiped out," Noah said.

As we drove the last hundred yards toward our compass points that morning, we despaired. There were a couple of ordinary brick ranch houses on one side of the road, spaced pretty far apart. Then we saw a dirt road that cut between them, marked with a simple wooden sign perched high on a pole with a single telling word painted on it (which I won't reveal in case a would-be poacher familiar with the area might be reading this book). We pulled off the pavement and onto the dirt road and parked beside a row of crape myrtle trees. No one was home, it seemed, at either residence. A mockingbird took the stage at the top of a myrtle to perform a complicated musical number. Otherwise, no signs of life. We waited, wondering if we were in the right place.

I was thinking about the most famous carnivorous plant, the Venus flytrap (*Dionaea muscipula*), which grows within a seventy-five-mile radius of the coastal city of Wilmington, North Carolina, and nowhere else on earth. Even after the state legislature stepped forward in 2005 to claim the herbaceous perennial to be the state's "official carnivorous plant," flytrap and pitcher plant poaching in the region became so frequent that wildlife officials tried marking the plants with an orange dye that would rub off on the poachers and leave the plants unharmed.

Then nine hundred flytraps went missing from North Carolina's Green Swamp Preserve, south of Wilmington, in a major heist in 2009. Subsequent incidents provoked the state legislature in 2014 to make flytrap poaching a felony in the five counties where the plants are most common.

A month after the legislation became law, a North Carolina man was arrested for stealing 970 flytraps from the Holly Shelter Game Lands, north of Wilmington. According to media reports, the twenty-three-year-old man and his three accomplices had dug up nearly 3 percent of the entire remaining population of flytraps in the area. A felony sentence and six to seventeen months of jailtime rendered to the driver of the vehicle was the first such judgment under the new law. The driver's accomplices, including his father, received probation. The plants were returned to the wild and reportedly survived.

A few months later, in November 2015, one thousand flytraps and two pitcher plants went missing from Orton Plantation—a preserve owned by a wealthy land conservationist in Wilmington. The perpetrators were arrested and held on $1 million bond because of prior convictions for wild plant poaching. This harsh crackdown appeared to create a deterrent for a few years until a repeat offender was recorded on a motion-activated game camera stealing plants. He received seventy-three felony counts for poaching 216 flytraps out of the Pinch Gut game lands in the Green Swamp in 2019. He could spend the rest of his life in prison under the new penalties.

Alabama does not have such stringent antipoaching enforcement, and I was a bit uneasy when, after about twenty minutes of waiting for our rendezvous with Chuck Byrd, a large, unmarked white truck slowed and turned into the drive behind us, blocking us in. Immediately the smell of diesel hit me. A workman of some sort—maybe for a utility company, I wondered? We were unnerved to be standing on private property with rare plants in the vicinity, not really knowing where we were or who owned the place and now blocked from the road.

The driver took his time getting out of the truck. Blue T-shirt, khaki pants, ball cap. When he closed the door and rounded the front of the vehicle, I could make out the words "Nature Conservancy" on his shirt. The cap also had the organization's name printed on it with one embroidered word underneath: "Caribbean." The cap had clearly absorbed some serious perspiration over time, perhaps in a tropical forest, I imagined. When Chuck Byrd took off his sunglasses, his eyes were the faded marine blue of his T-shirt. He had a friendly, open smile and a salt-and-pepper beard.

We made quick introductions, and Chuck explained that his smelly vehicle was actually an F-6 fire truck that he drives on days when he is managing controlled burns on conservation lands. The burns, generally sponsored by Fish and Wildlife or by contract with a private landowner, are Byrd's main activity this time of year. The Nature Conservancy in Alabama provides the service because it is essential to the vitality of many species of plants and trees, including the Alabama canebrake pitcher plant. The forest we would enter, Chuck said, was burned the previous year and was on a two- or three-year cycle of burning, regeneration, and maintenance. Burning during plant dormancy clears away pine needles, dried weeds, and grasses and allows the soil to be exposed again to rain and the seeds that follow after spring flowers bloom. Pitcher plants can reproduce by extending roots and shoots (called rhizomes), and they will also drop seeds that are about the size of a poppy's. If the ground cover is too thick, the tiny seeds have a hard time gaining purchase in the soil to sprout, Chuck explained. He was clearly proud of his white fire truck, and there was also a little spark in his eye as he talked about the adventure of setting fires in the woods.

"We did a burn yesterday," he said. "We're almost at the end of the season. It runs from December/January to May/June. I went to work yesterday at four in the morning and got home at eight last night." He shook his head. "I got about five hours of sleep, so I'm a little foggy."

Chuck climbed back in his truck and drove ahead of us to lead the way down the rutted road beyond the ranch houses, down a hill, and into the forest. He pulled up to a gate, unlocked it, and shoved the heavy metal barrier out of the way. We parked in a grassy area a few yards ahead and he locked us back in. Mature hardwoods around us gave some shade.

Chuck Byrd, a land steward at The Nature Conservancy in Alabama, conducts controlled burns across the state to benefit endangered plant species, including several species of pitcher plants.

Hunters are allowed in this area during a short window each winter, Chuck said, which is good public relations for The Nature Conservancy.

The ground was sandy white. It was already warming up, but not hot yet. A slight breeze helped. Chuck had suggested in advance that we wear long pants, long sleeves, and boots, and Donna had already cheerfully reminded me that we had seen several dead snakes that morning, along with a number of armadillo roadkills. "It means that the environment here is healthy," she said. "We used to see dead snakes on the road all the time when I was a child in North Carolina, but not so much anymore," she added helpfully.

What we forgot to bring was our insect repellent—a DEET-free and pleasantly scented lemon and eucalyptus oil that claimed on its label to repel "mosquitoes that may carry Zika virus, West Nile virus, and Dengue virus." Oh, well.

We set out straightaway on foot down the rock-littered dirt road. Young longleaf pines on either side of us looked rather like animals with long glossy fur, arms outstretched, and surrounded by knee-high yellow grasses. Instead of heads, though, the pines had amazingly long, yellow pollen candles at their tops, standing up bright against a dense blue sky.

I asked Chuck how he got into this work. An Alabama native, Chuck lives with his wife and son in Alabaster, a town named for one of the varieties of marble found in the region. (Marble is a metamorphic rock that has been transformed from the great flanks of sedimentary limestone also underfoot in this part of Alabama and key to its distinctive plant populations.)

After earning a forestry degree from Auburn University, Chuck worked in the paper industry in his first job. Now forty-four, he has worked for The Nature Conservancy for nine years. "I am amazingly lucky," he said. "[The Conservancy] has opened up so many possibilities for me." He has performed conservation management in a variety of ecosystems and built canoe launches on the Cahaba River (where the lilies described in chapter 5 grow). These days he works as far north as Huntsville and covers the entire state of Alabama in fire season.

We stopped on a ridge for Chuck to get his bearings in the forest. "One thing that has helped TNC do this work in the South is people's connection to the land," he said. "Other places that have not had a cultural tradition of fire have not thrived. Farmers have burned their fields here for centuries, starting with the Indians. That really helps."

Chuck explained that there were actually six sites with pitcher plants within a twenty-mile radius of where we were standing. "This is a seepage slope that has an impervious rock layer just under the soil," he said. "The rock is acidic. Rain soaks in, hits the impervious layer, and shoots downhill." That landscape is what canebrake pitcher plants prefer. The "canebrake" in the plant's name refers to the giant native bamboo that once thrived in dense wetland stands throughout the South (to be discussed in chapter 8).

Populations of pitcher plants were destroyed over the years by railroad and pond construction and gravel mining that also altered the natural water table. "Lots of pitcher plants grow in areas where water is close to the surface of the ground, so if you are a landowner wanting water for your cattle, you tend to dig out the wet spots. That's part of what has happened to the green pitcher plants over time," Chuck said. For the canebrake plants, destruction came in the conversion of pine savannahs to pine plantations, which was what had come to pass for a time on the acreage where we were standing.

"This property was commercially planted in timber," Chuck said. "TNC bought it through our land acquisition program called Forever Wild. We did a commercial harvest, marking plants in no-take areas, and then we took out lots of trees to open up the forest floor to sunlight. Trees also suck up the moisture, and other plants can't grow when the pines are commercially planted so close together as they once were here."

Amendment 543 to the Alabama Constitution, which created "Forever Wild Alabama," got the highest vote among bills passed in 1992, when the legislation was brought forward. It was then approved later in the year by 84 percent of Alabama voters in a public referendum. As the state's longtime conservationist and writer Pat Byington put it, "In 1992, Alabama had the least amount of public land set aside for conservation and wildlife in the South. The state of Alabama had no plan or programs to expand parks, nature preserves and wildlife areas." By 2019, more than 266,000 acres had been acquired since the public resoundingly approved the referendum.

The movement to create the legislation began when a group of thirty-three citizens from divergent organizations and businesses were tapped by the state to devise a plan. At the group's first meeting, a facilitator asked everyone to list their hopes on one sheet of paper and their fears about the process on a second sheet. Each then shared their lists by reading aloud. Kathy Stiles Freeland, the first director of the Alabama chapter of The Nature Conservancy, later described the moment: "We sat and looked at each other. We had a common bond: the same hopes and fears. And that made us begin to see each other as people, other than an organizational representative and thus, the enemy."

In a growing spirit of trust and collaboration built through a series of early morning meetings at state park lodges across Alabama, the group eventually drafted the legislation. "The public vote felt very good," Chuck Byrd said. "It was such an affirmation of our work. People in Alabama love their woods, but we get a bad rap in the national news. And some of it is deserved. We are still battling ATVers" (all-terrain vehicle drivers).

Chuck went on: "I think the older generation has a strong land ethic. Younger folks weren't taught this in school, but many have become self-taught, which means they have a drive to learn. Now there is even a Face-book page for Alabama's rare plants, so there are better means for connectivity among those who share an enthusiasm for plants. We have a history of hunting and fishing here. People pick up on this. A lot of people really love Alabama and the plants and wildlife, so they want to help conservation. The longer I live here, the more hope I have."

Chuck moved on ahead of us, down a slope. "Here they are," he said finally. At first, I didn't see them—the pitchers and their unusual red blooms on stalks above them. The blooms put me in mind of pinwheels—those triangular curls of paper pinned to a stick that I learned to make in grade school.

"I feel like Godzilla walking through Tokyo here," Chuck said, nearly stumbling on the slope. "This band of specimens is maybe a hundred yards long by twenty yards wide, and I could stomp all over them if I didn't know how to look for them." Plant blindness, I thought. I probably would have missed them without Chuck leading the way.

The pitchers were smaller than expected—delicate, translucent, with red veins running up and down the yellow throats. Chuck pointed out tiny sundew plants (*Dropsera*), also carnivorous, that had settled themselves under the pitcher plants. "The sundews have super-tiny hairs that collect gnats or smaller bugs for food," he said.

Then I noticed the full flank of plants that stretched along the ridge before us, surrounded by low shrubs that looked familiar.

"Wild blueberries," Chuck said as I raked my hand over a bush. "*Vaccinium angustifolium*."

I looked around, trying to take in the whole ecosystem, noticing the pines that also occupied the sloping ridge. They were spread out at some distance from one another. "These pines look young to me," I said. "They have skinny trunks."

"Remember, the soils here are nutrient-poor," Chuck said. "Some of these trees might be a hundred years old, but they are not well fed."

Donna was shooting images of the pitcher plants from all angles.

These rare Alabama canebrake pitcher plants (Sarracenia alabamensis) bloomed at an undisclosed location in the wild under the close observation of The Nature Conservancy in Alabama.

"I would love to know how those Michigan botanists found these populations of plants in the first place," Chuck said, "back when the commercial forest was here." He went on to explain that The Nature Conservancy has regularly taken seed from the plants and given them to ABG to grow their own specimens. "Unlike the species of pitcher plants that thrive in a bog, these grow on this seepage slope in a pretty narrow window. Eventually their color changes and they die back, but then they come back in late winter, which is not a very hard winter here." Chuck bent down on one knee and lifted a flower head so that we could see inside the pinwheel.

"These seedpods dry out." He pointed inside the bloom, where tiny envelopes of seeds were tucked. "The seeds fall into the blossom cup to be dispersed by wind and rain. Poachers like to talk about the hood shape as a feature on pitcher plants." Chuck pointed to the way the cap came up and hovered over the opening to the throat. Unlike the Venus flytrap, this hood does not close up and trap insects that venture into the pitcher. Instead, the hood serves as a kind of festive-looking umbrella against rainfall. Keeping out some of the rainwater ensures that there is no dilution of the juices that digest the plant's dinner.

Aaron Ellison, a senior research fellow at Harvard, has written that people have long understood the pharmacological uses of carnivorous plants. Native Americans used extractions from pitcher plant leaves to treat diabetes; extracts from sundews were used to treat respiratory ailments, including tuberculosis, as early as the mid-1500s. The roots of yellow pitcher plants were also used in the treatment of indigestion and headaches and were studied at length by a South Carolina botanist in the 1800s. Researchers are investigating the fluids that pitcher plants use to trap and digest insects in the development of biomimetic materials that may lead to "new drug-delivery systems, self-lubricating machines, and new types of agricultural drip-irrigation systems," Ellison wrote. These wide-ranging applications are a strong argument for preserving these plants in the wild, because non-native plantings of carnivorous plants can become invasive or disturb the unique genetic identity of their neighbors and lead to the demise of what makes each species unique.

"My son thinks these plants are cool just because they eat bugs," Chuck said. We laughed. "You are looking at the third-largest site for wild pitcher plants in the state, so we are careful who we tell about this place." Alabama is number five in total biodiversity in the nation and number one in pitcher plants. A wave of gratitude for the moment washed over me.

As we had learned before coming to this forest, there are a lot of listed plants and candidate plants for the federal government's endangered designation, but not a lot of funds for protection. "It's hard to make a case with the feds for help against poachers," Chuck explained, "and the State of Alabama means well, but they don't have the manpower to catch people."

Donna took a few more photographs, then we headed back up the slope. Chuck told us how a teacher had recently taken a group of students to Little River Canyon: a US Preserve managed by the National Park Service on top of Lookout Mountain in northeastern Alabama, near Fort Payne. "She showed her students a patch of the green pitcher plants in the wild. Four days later, all the plants had been dug up," Chuck said, shaking his head.

As we headed back to our vehicles, an arresting little plant close to the ground caught my eye. It was orange and puffy looking. "What's that?" I asked, and without missing a beat, Chuck said, "That's a bog Cheeto." He slapped a mosquito on his neck. We laughed. "Honestly, I can't remember the botanical name at the moment. You'll have to ask my boss, who is the botanist on our team." (It was *Polygala lutea*, I later learned from Noah Yawn.)

The Nature Conservancy has a staff of some twenty people covering the whole of Alabama. In fire season, Chuck says, they enlist assistance from the Sierra Club and the Audubon Society to help with prescribed fires. In December, they burn for the green pitcher plants. "I tell people I am a plant conservationist, but I set them on fire, and I mow them down." Chuck smiled. He explained that to prepare for a burn, they will cut away the understory and use a leaf blower to gently create fire breaks in the forest. "If you disturb the dirt too much, you invite invasives," he said.

Next we came upon a flourishing spread of bracken fern. "First thing to come back after a burn are those ferns," Chuck said. "The burn causes a flush of nutrients, and the green goes wild." Just then, as we had nearly made our way back to the road cut, Chuck spotted something else. "Canebrake rattler," he said, and I jumped. But it was only a shed skin curled around some rocks. "They are shy," he said. "We usually only see them when we're burning, and then they start moving."

When we arrived at our starting place, I asked Chuck if there was anything else he thought people ought to know.

"I want people to understand that Alabama is a unique state, and while you may not always see the great biodiversity, it is here. A healthy environment is great for everyone. Tender vegetation is great for deer. These grasses around us have a high percentage of protein. Though I am a forester by training," Chuck said, looking up into the trees around us, "the really important stuff is going on from the waist down." He pointed to the dirt. "That's what I tell people when I give talks around the state. The Southeastern Grasslands Initiative [a twenty-three-state conservation project based at Austin Peay State University, in Tennessee] found that in a square meter of the Southeast, you may find ten to twenty different species of plants. That is just amazing."

The next day, we drove to Little River Canyon National Preserve, which stretches out to the northeast, near the border of Georgia and not far from Chattanooga, Tennessee. More than fifteen thousand acres flank the Little River there. The canyon was designated a national park in 1992 and features one of the deepest gorge systems in the Southeast. Sandstone cliffs and bluffs tower over the river below, where the whitewater chutes are a challenge, even to expert kayakers. The green pitcher plants in this area are the only ones growing in Alabama. Located at the southern end of the Appalachian chain and above the fall line, they've been on the federal government's endangered list a decade longer than the canebrake pitcher plants.

At the park's visitor center, a building made of handsome local rock and timbers, we approached the staff asking where we might be able to find the pitchers. The ranger hedged at first and then finally, if vaguely, suggested a turnout "down the road a ways," on the lip of the canyon where the landscape was regularly mowed. When we arrived, we found a stunning overlook into the canyon, with enormous boulders and stomach-flipping ledges that, to our relief, were equipped with guardrails. Though we hiked around the area, we didn't see any pitcher plants.

We then traveled farther north, to historic DeSoto State Park, perched on Lookout Mountain about eight miles northeast of the town of Fort Payne. Developed by the Civilian Conservation Corps in the 1930s, this park offers campsites, quaint cabins, and miles of hiking trails. The rangers supply hikers with a handout listing wild and rare plants, especially noting the presence of the delicate pink lady slipper, Alabama's largest native orchid. The brochure warns visitors: "Take only pictures. Leave only footprints. Kill only time." As it turned out, we didn't need to ask where to find the green pitcher plants. Just outside the visitor center, a magnificent display of *Sarracenia oreophila* had been created in 2013 by Austin Evans, a local Boy Scout working toward his Eagle designation.

"We plan for this display garden to be permanent," the park's naturalist, Brittney Hughes, told me later. "The plants came from our own seed stock gathered here in the wild. They were grown at the Atlanta Botanical Garden and brought back here."

The green pitcher plant was first documented in 1900. It now grows mostly on private property across four Alabama counties. Mountainous spots in Georgia and North Carolina also have a single population each that Fish and Wildlife officers in those states are charged with protecting. The specimen garden at DeSoto featured grand, translucent flutes of varying heights and angles emerging from the soil, which somehow put me in mind of Grumpy's pipe organ in the classic cartoon animation of Disney's *Snow White*. I could have stared at the luminous pitchers for hours.

"In my experience, the landowners who have pitchers on their property were some of the nicest people I've met," Noah Yawn told me six months later, while he was writing up his Alabama census data. "They are folks who just happen to have these extremely rare plants on their land. Some were worried that the government might take their land because of the plants," he explained. "I realized I needed to help them understand that wasn't going to happen and to show them just how special the plants are."

Noah said at one site noted on the 1995 census, they had found the pitcher plants again, but they were in very bad shape. The owners didn't know anything about them. When Noah pointed out the pitchers, they said, "Well, what's so special about this?"

Emily Coffey, vice president of conservation at ABG, arranged to send the landowners tickets to visit the Atlanta garden so they could see the robust descendants of their plants, grown from seeds taken from their property. "Big, beautiful lime-green plants," Noah said. "They were stunned. They are gung-ho now about preserving them."

Noah also told us that six months after our foray with Chuck Byrd, someone had come to the secret conservation site and taken every single seedpod from the population of canebrake pitchers we had seen in May. Another population nearby, planted in the wild from seed cultivated at ABG, was also stripped of seed. The second group of plants, Noah said, were even healthier than the ones we saw—taller and more saturated with color.

Then, sometime later the green pitcher plant sites in northern Alabama had their seeds removed. In North Carolina, seeds also went missing from the extremely rare mountain sweet pitcher plant. "Does someone have a perverse conservation agenda?" Noah wondered out loud. He'd been keeping an eye out for any discussion of seeds on the various sites on Facebook where carnivorous-plant fans share information.

At ABG, I asked Emily Coffey what she thought of the disappearance of the seeds in the wild. She smiled and said, "Well, that was better than poachers grubbing up the plants. This thief was someone who was at least informed. I have to take that as an improvement." The plants would make more seeds the next year, she said.

On a rainy Monday in February when ABG was closed to visitors, Emily led us through a maze of exhibitions to enter the behind-the-scenes greenhouses where the garden was cultivating a variety of endangered species. (ABG does not offer plant sales, as many other gardens do. Instead, they use their greenhouses strictly for conservation cultivation.)

Emily, an attentive listener and rapid-fire speaker, has flaming red curls that bounce around her face. She grew up in the radiance of the world-class Missouri Botanical Garden in St. Louis; earned her master's and PhD degrees in long-term ecology in England at the University of Oxford; and then taught biology at the University of North Carolina–Asheville. She took her present position at ABG in 2017, leading the garden's endangered plant research and directing the Southeastern Center for Conservation. It's a big job, but she seems indefatigable.

As we toured ABG's newest facilities, it was clear that Emily's passion and diligence are contagious among staff and volunteers who work in the exhibitions and behind the scenes—pulling weeds, watering, dividing and repotting specimens, and minding the minuscule indicators of each plant's health. Emily stopped us to point out the rarest of ghost orchids and then again a bit farther down a path to admire a huge, exotic lady slipper that seemed to be wearing red dreadlocks.

Passing through the tropical plant collection, she introduced us to ABG's resident alligator snapping turtle, hanging out in a baby pool while its much larger waterfall pond was being scrubbed. Then we ran into our fellow Florida Torreya adventurer Trey Fletcher and got a quick look at the collection of exotic succulents that he manages.

When at last we stepped into a humid glass greenhouse and Donna's camera lens fogged up, the expanse of plants seemed overwhelming—tray upon tray and row upon row of pitchers of many colors, shapes, and heights. The trays were set out on waist-high tables that seemed to run the length of a football field. Emily introduced us to the pitcher plant curator, John Evans, who pointed out various species, including specimen trays of the Alabama canebrake. Then he led us into a back room where tiny flutes of baby pitchers warmed under grow lights.

John explained how challenging it can be to grow young pitcher plants in a greenhouse, where they can't reliably feed on their usual diet of insects. Given the pitcher plant's preference for nutrient-poor soil, fertilizing them in this situation requires a light touch. Eventually, the plants would be ready to move into larger trays for the greenhouse, he said, and many will go later to ABG's outdoor nursery, on the north side of Atlanta in Gainesville—a facility not open to the public. Rather, the site is part of ABG's safeguarding program for endangered species. More than a half-dozen pitcher plant species are grown there, coded by their geographic origin and specific population. Each subspecies is kept at a distance from others, to avoid hybridization. Not only do these cultivated plants help preserve species numbers, but many pitchers have also been repopulated in the wild—especially the Alabama canebrake pitcher plant—to ameliorate its radically reduced numbers.

The Georgia Plant Conservation Alliance has led the way among southern states in formally articulating best practices and protocols for safeguarding species, including seed collecting and plant propagation at botanical gardens. The group is helping other states get on board with the same practices.

ABG's Conservation Safeguarding Nursery in Gainesville protects some of the wild South's rarest and most endangered natives and meticulously conserves variety in the gene pool. Both the Atlanta and Gainesville sites are also training the next generation of plant conservationists: the top undergraduates and graduates that they can find in the region and beyond. Emily Coffey has led an effort at ABG to enhance the representation of women, people of color, and low-income students through paid internships with the garden. With grant funding, ABG has also been able to offer memberships to a broader range of income groups in the city and to make a dent in the dominance of Euro-Americans in the environmental field. ABG was founded by a group of white ladies who belonged to a rather exclusive garden club on the north side of town, but times have changed in Atlanta, and ABG is leading the effort to broaden the diversity of groups engaged in plant conservation.

As our tour ended, we thanked Emily for her time and vowed to return to ABG when we could walk among the thirty acres of outdoor gardens without an umbrella and once again visit with our new friends from the Florida Torreya expedition.

Our final destination to observe an amazing array of pitcher plants from throughout the South was Auburn University in Auburn, Alabama. The Donald E. Davis Arboretum on campus is one of ABG's most active pitcher plant conservation partners in the region. Curator Patrick Thompson has worked for more than twenty years at the arboretum and now directs the

Emily Coffey, vice president of conservation at the Atlanta Botanical Garden (ABG), examines a tray of pitcher plant seedlings (Sarracenia alabamensis) emerging in an ABG greenhouse.

Alabama Plant Conservation Alliance, which the Georgia alliance helped to launch.

Patrick, a bearded man most often found in a T-shirt and khakis directing his team as they prune and mulch the manicured grounds, told us that his interest in the natural world began with a pet turtle. By the time he was ten, he said, he'd joined the Alabama Herpetological Society and thought he wanted to study snakes and amphibians for the rest of his life. Once he got to college at Auburn, however, he developed a fascination for plants and their essential role in evolution. Now he is safeguarding endangered plants in the 13.5-acre facility in the heart of campus and curating a collection of a thousand taxa, of which only about a dozen are not native to Alabama.

"Our mission," he said, "is simple: to inspire people to appreciate plants." He added that Alabama has 3,500 distinct plant species. The arboretum's three core collections feature twenty-five native species of Alabama azaleas; a collection of oaks, including the rare Boynton oak; and a colorful pitcher plant complex alongside the arboretum's designated outdoor classroom.

Patrick joked that working at the arboretum gave him permission to indulge his hoarding tendencies. "This is a safe place for me to collect things," he said. In his spare time, he cares for a bonsai collection at home, including a sixty-year-old evergreen that he has babied since it was passed along to him by an elder. "Yes, I also torture plants at home in the Japanese horticultural tradition," he said, deadpan. We laughed.

Patrick was quick to tell us how much he has benefited from the enthusiasm and dedication of his employee Noah Yawn, who soon joined us for a tour of the arboretum. Heading toward his senior year, Noah was trying to figure out his plans for graduate school and how he can make a living doing what he loves most. He works two days a week for his mentor, Patrick, and the two self-described plant nerds also spend many weekends off-campus using chain saws to clear the way for sunlight to reach seepage bogs in both private and public pitcher plant preserves, where such maintenance is welcomed.

Noah admitted that college for him had been academically lonely. He was the only undergraduate focused on botany at Auburn, he said. His major—Organismal Integrative Biology—was created to combine many facets of biology into a single rigorous course of study that can also include a conservation focus. Meanwhile, Auburn's botany program was phased out. The university's outstanding botany faculty retired, and individual members were not replaced. "There are lots of plant people here,"

Organismal biology student Noah Yawn and arboretum specialist Patrick Thompson stroll through the fourteen-acre Donald Davis Arboretum on the campus of Auburn University in Auburn, Alabama.

Noah said, "but they are not necessarily interested in the workings of bio-diversity and botany. They are more focused on landscaping or maximiz-ing and optimizing agricultural production."

As a first-generation college student from a low-income family, Noah has done very well in his studies. He is also pursuing a second major in geology and is increasingly confident that he can find a niche in conserva-tion work because of the growing shortage of botanists. He worries more about what he calls "the chosen path of ignorance" that he sees among his peers. "It's more than plant blindness," he says. "Their whole priority seems to be shifting away from the natural world. Many students studying biology are pre-med. They will likely never be exposed to biodiversity be-yond their introductory biology courses taken freshman year, which even then rarely cover biodiversity and conservation, let alone Alabama's own backyard of plant riches."

Patrick affirmed Noah's concern. He said that botany as a major fizzled out as the natural sciences became more focused on pre-med students. And such a shift is not confined to Auburn, he explained. "We are going to run out of botanists right in the middle of this biodiversity crisis. With horticulture programs closing across the country, we could really be in trouble. Working with plants is not a job that can be replaced by robots."

Still, Noah gives Patrick hope, he said. As we strolled through the arboretum in the late afternoon light, the pink and red azaleas were already blooming, and pollen candles were swelling on the pines. We were headed for the pitcher plant bog and outdoor classroom—a fantastic display that we had already seen at its peak in May. Noah took the lead, while Patrick hung back on the trail. "You know, when Noah was still in high school," Patrick said quietly to me, "he posted a request on a national website for pitcher plant enthusiasts. He told the group that they should donate to the Davis Arboretum to help our pitcher collection, and people actually did it!" Patrick pointed to a greenhouse ahead of us. "The collectors paid for this building so that when we gather seeds from Alabama's rarest pitcher plants, we have a place to grow them."

When we came to the fenced area behind the outdoor classroom, Noah walked over to an array of black plastic trays that had been set outdoors to take in the sun on this warm afternoon. The trays were filled with green pitcher plants that Noah and Patrick had cultivated from seed collected at DeSoto State Park. They would be used for outplanting in the wild as part of the Auburn Arboretum's safeguarding and restoration work with the Alabama Plant Conservation Alliance. The lime-green plants were no more than two inches high and crowded together, neck and neck. Noah ran his hand across the tops of the pitchers, and then he and Patrick each picked up a tray so Donna could shoot a little video. Noah looked at the camera. "There are more *Sarracenia oreophila* here in this tray than still exist in the wild at the site they are from," he said. "For the species as a whole, only six thousand individuals remain. Think about that. An entire species represented only by six thousand plants left in the wild."

4

Miccosukee Gooseberry

Perhaps the most remarkable thing about the Miccosukee gooseberry (*Ribes echinellum*) is that it survives at all. This deciduous, perennial shrub, which can grow from three to four feet tall, is a member of the currant family and is globally classified as critically imperiled. Only two wild and somewhat scattered populations are known (on private lands in Florida in the vicinity of Florida's Lake Miccosukee and in South Carolina on public lands in the Sumter National Forest), making this gooseberry one of the rarest plants in North America. The gooseberry's story suggests that at one time there might have been many more species in Florida that we will never know about, plants lost to residential development, commercial enterprise, and agriculture, long before botanists could identify them.

I first saw this gooseberry plant in the Steve Church Endangered Species Garden at the Sarah P. Duke Gardens, at Duke. The leaves looked like those of a wild geranium to me, only larger and deeply embossed. The greenery obscures some disagreeable half-inch thorns that are spaced along the gangly limbs of the shrub, ready to stab the unknowing passerby.

Oddly, Miccosukee gooseberry sheds its leaves in summer, and they begin to grow back in fall. In spring, delicate, pendant flowers emerge along the stems, which I was able to see on the lone specimen in the North Carolina Botanical Garden in Chapel Hill. The South Carolina naturalist Patrick McMillan has called the blooms "dainty, dangling ballerinas"—an accurate description. From them, spiny, inch-diameter fruits emerge that look something like bright green miniature sea urchins.

Research suggests that two species of bees—both of which are present in Florida and South Carolina—are the plant's main pollinators. The botanists who first documented the species, in Florida in the 1920s, believed

When the Miccosukee gooseberry (Ribes echinellum) is in its native habitat, the delicate pendant blossoms will eventually become spiny fruit. This specimen, conserved in the collection of the University of North Carolina Botanical Garden in Chapel Hill, is growing nearly three hundred miles northeast of the plant's native habitat in South Carolina and is not likely to form viable fruit.

that it might hold some commercial promise as a fruit crop. The South Carolina population was not discovered until thirty years later, in the 1950s. Why the Miccosukee gooseberry has been present in these two distant, but geologically similar, sites has been a great, unsolved mystery for botanists. Complicated hypotheses abound.

Both the South Carolina and Florida locations are rich in species that are rare elsewhere, and both have a highly acidic soil that stays moderately wet. However, other gooseberries in the *Ribes* family—red currants and black currants, in particular—tend to grow at much higher elevations or farther north. Did this species range more widely across the South at one time? No one really knows.

Some biologists have speculated that this gooseberry goes back more than ten thousand years, to the Pleistocene era, the last glacial period on earth, when plants like the Torreya tree were pushed into northern Florida beyond the glacial edge. Some hypothesize that the plant depended on large mammals that grazed on midstory shrubs, keeping its habitat open to sunlight. Others have suggested that birds and perhaps other mammals ate the berries and carried their seeds across the landscape. (Seeds that pass through the digestive tracts of animals can be "scarified," or roughed up, so that when they are eliminated, they more easily take root in the ground.) One premise is that the extinctions of the passenger pigeon and the Carolina parakeet may have contributed to the loss of the plant's range. The story of these two most unusual birds has often been told in ecological literature and is worth repeating here.

The Smithsonian Institution estimated that there were 3 billion to 5 billion passenger pigeons on the continent when Europeans arrived in North America. Long before British colonial expansion took root and official borders between the American colonies were drawn on paper, the birds migrated yearly from north to south in great flocks. According to the Canadian biologist Paul M. Catling, they eventually came to be called passenger pigeons because the traveling flocks looked like long passenger trains crossing the sky. "The droppings fell like hail," Catling added. The populations of these gregarious nomads were so vast that witnesses said they blotted out the sky and eclipsed the sun. Come winter, the birds traveled to roost in the mixed hardwood forests of the wild South—from what is now North Carolina to points as far south as the Gulf Coast.

Until their extinction at the beginning of the twentieth century, passenger pigeons created what must have seemed like an apocalyptic spectacle, speeding along at nearly sixty miles an hour, thrusting forward in wide battalions, hundreds of thousands of wings clapping with a sound I would imagine was greater than any burst of applause emanating from a contemporary sports arena.

When the birds descended at evening to roost, they simply piled on top of one another in trees. So great was their number that early settlers reported that limbs often cracked and broke under the birds' collective weight, creating a nightlong racket.

"Frequently beneath the roosts, sometimes forty miles long and three miles wide, dung accumulated to a foot deep, and killed the understory vegetation as well as all the trees," Catling wrote. Though it sounds repellent, the bird droppings contributed to soil conditions that would have been favorable to calcium-loving plants, including the Miccosukee gooseberry. As Catling put it, "The southeastern woodlands would have been more fertile and had higher calcium levels due to the fertilization by huge flocks of both passenger pigeons and Carolina parakeets."

Carolina parakeets—the only parrot native to North America—also thrived in the wild South, particularly in cypress swamps and bottomlands. Because they ate seeds, fruits, and nuts, they became an agricultural menace to European settlers, and farmers shot them in great numbers. Soon their large, colorful feathers were in great demand for ladies' hats. While their habitat was increasingly disrupted, the birds also fell prey to specimen collecting for taxidermy and the caged bird trade. The last sighting of the Carolina parakeet in the wild was in Florida in 1904, and the last known captive, named Inca, died in 1918 at the Cincinnati Zoo.

It was precisely because of passenger pigeons' tight-knit, communal nature that they, too, would be eradicated. In a thirty-year span before the beginning of the twentieth century, English immigrants not only shot the birds in improbable quantities for food but also interrupted their nesting habits through the relentless momentum of forest removal and human settlement. The last passenger pigeon kept alive in captivity also died in the Cincinnati Zoo, at the age of twenty-nine on 1 September 1914. Her name was Martha, for Martha Washington. In 1947 the eminent environmentalist Aldo Leopold poetically reflected on the bird's demise in an essay he read at the dedication of a monument to the species:

> The Passenger Pigeon was no mere bird, he was a biological storm. He was the lightning that played between two biotic poles of intolerable intensity: the fat of the land and his own zest for living. Yearly the feathered tempest roared up, down, and across the continent, sucking up the laden fruits of forest and prairie, burning them in a travelling blast of life. Like any other chain reaction, the pigeon could survive no diminution of his own furious intensity. Once the pigeoners had subtracted from his numbers, and once the settlers had chopped gaps in the continuity of his fuel, his flame guttered out with hardly a sputter or even a wisp of smoke.

If Miccosukee gooseberries were among the "laden fruits" of the forest that the passenger pigeon and Carolina parakeet consumed, it is hardly a leap to recognize in Leopold's words how humankind was also *a biological storm, a traveling blast of life*, particularly in Florida, where the gooseberry was discovered and named, a decade after Martha expired. That human awareness of the existence of this particular gooseberry came so late makes it difficult to connect it definitively to these birds who might have provided a seed dispersal mechanism. Still, the relationship is plausible.

Because of its semitropical climate and unusual soil chemistry, the Florida Panhandle was a magnet to botanists even before Florida's statehood was established. Hardy Croom, who first documented the Florida Torreya tree, as discussed in chapter 2, was one of the early explorers in the vicinity of the gooseberry, but if he saw it, he took no note of it. Croom was a North Carolina tobacco plantation owner and one-time state senator who, along with his brother, inherited 2,400 acres near Tallahassee from their father.

Croom traveled with his brother to Florida for the first time in 1831 to inspect their inheritance. Though skirmishes with the Indigenous tribes were an ongoing concern for settlers, Croom was smitten with the landscape. The prospect of botanizing in the area was irresistible to him, and he began plans to build a grand estate there for his young family. The property, called Goodwood, is now on the National Register of Historic Places. The buildings, gardens, and museum are only about nineteen miles as the crow flies from the Florida populations of gooseberries on the east side of Lake Miccosukee.

Croom abandoned his tobacco plantation in New Bern, North Carolina, resigned his state senate seat, and moved his enslaved labor force to Florida to begin growing cotton and corn at Goodwood. He also leased a second plantation in 1832 on the west bank of the Apalachicola River, and it was on this land that he discovered and named several native plant species new to science.

In addition to *Torreya taxifolia*, which the locals called stinking cedar, Croom found the Florida yew tree, *Taxus floridana*, which is also now extremely rare and confined to the area around Torreya State Park. Botanists are quick to point out that the eventual value of such finds is often unimaginable at the time of their discovery. In the case of the Florida yew and its relative, the Pacific yew (*Taxus brevifolia*), chemists later identified a compound in the tree bark that was an effective treatment for several forms of cancer. The compound was first tested in 1962, and after thirty years of trials, the FDA approved a drug called Taxol in 1992 for treatment of ovarian cancer. It was also approved for the treatment of breast cancer two years later.

The Florida species of the tree by that time was much too limited in numbers to provide raw material for drug production. During the drug's development, conservationists also worried that the Pacific yew might be overharvested. In response, Professor Robert Holton of Florida State University devised a synthetic formulation dependent on the more common yew sold in the nursery trade. In 2000, Taxol became the best-selling cancer drug ever manufactured.

Hardy Croom was also credited with the discovery of a small perennial herb with heart-shaped leaves and tiny nodding flowers that combine green petals and unusual maroon stamens tipped with orange. Croom found the plant in the shade of the Torreya tree, and his correspondent and mentor, the New York botanist John Torrey, named it Croomia (*Croomia pauciflora*) in his honor. *Croomia* is endangered in Florida now too, owing to its destruction by feral pigs. Through these finds, Croom was building his reputation as an amateur botanist. He published papers in the *American Journal of Science* on his findings. Then tragedy struck.

Though he had already relocated his business interests away from North Carolina, Croom struggled to convince his wife, Frances, to move to the wilds of Florida with him. While her husband worked in the Deep South, Frances spent most of her time in Saratoga, New York, and New York City. Their children—ages fifteen, ten, and seven—were enrolled in top private academies in the North. Frances told her husband that she was afraid for their safety in Florida.

The couple finally compromised: they agreed on a plan to move to Charleston, South Carolina, though Croom was still quietly preparing Goodwood as a family residence. In October 1837, Hardy Croom, his family, and Frances's aunt boarded a paddle steamer in New York City bound for Charleston with some ninety passengers and a crew of forty. Two days into the journey south, heavy seas from a hurricane caused the boat to spring multiple leaks. The crew and passengers attempted to bail the rising water as darkness fell, but the steamer, which had been engineered for river travel and not the ocean, broke apart after running aground a quarter mile off the coast of Cape Hatteras, North Carolina. None of the Crooms were among the thirty-nine terrified survivors, who were soon rescued and taken to Ocracoke Island. Eventually, the bodies of Frances Croom and ten-year-old William washed up on shore. The bodies of the other family members were never found. On the day of his death, Hardy Croom had turned forty.

Another amateur botanist in the nineteenth century, Dr. Alvan Wentworth Chapman, also lived within striking distance of the Miccosukee gooseberry population in Florida without finding it. Originally from Massachusetts, Chapman settled in Quincy, Florida, to practice medicine in the vicinity of what is now Torreya State Park. He met and befriended Hardy

Croom when the younger man leased his second plantation, near Quincy. Chapman was inspired by Croom's discoveries. The two had planned to botanize the Panhandle together before Croom's untimely death.

Chapman later moved his practice to the Gulf Coast at the mouth of the Apalachicola River, where he quietly supported the Union during the Civil War by treating injured soldiers during their blockade of the port at Apalachicola. By all accounts, Chapman was a brilliant man who spoke and read several languages and often traveled into the Florida frontier at some risk to care for his patients. He collected native specimens on these ventures and claimed in a letter to John Torrey that his horse knew the local plants better than he did.

Sometime before 1865, Chapman found and documented the wild Florida azalea (*Rhododendron austrinum*), which is sometimes called the honeysuckle bush for its orange/yellow color and lemony fragrance. This shrub is abundant in northwestern Florida and is increasingly available in the horticulture trade because of its excellent features. Chapman's monumental regional compendium, *Flora of the Southeastern United States*, was published in 1860, and was the standard reference on the topic during his lifetime. It was not supplanted until 1903, when John Kunkel Small's inventory with the same title was published.

In her thorough profiles of early Florida botanists collected in *Journeys through Paradise*, the environmental writer Gail Fishman explains that Chapman suffered from color blindness and sometimes had difficulty identifying plants. His most celebrated discovery was Florida's only native rhododendron, found on the same bluffs as the Torreya tree.

In 1875, Chapman and his wife hosted the Harvard botanist Asa Gray when he came to Florida, and it was Gray who confirmed the uniqueness of the rhododendron species by its unusual pinkish color, which Chapman's eyes could not discern. Gray saw to it that the rhododendron was named in Chapman's honor: *Rhododendron chapmanii*.

Chapman died in 1899, just short of his ninetieth birthday and only days after he had been out hunting local plants to send to Harvard's Arnold Arboretum. He is buried in Chestnut Cemetery, just behind his home, which still stands in Apalachicola. The rhododendron that bears his name is now listed as critically endangered.

In correspondence, I asked Gail Fishman for her thoughts about why neither Croom nor Chapman, who surely came close to Lake Miccosukee in their explorations, had found the local gooseberry.

"Even though Croom lived for a short time on Lake Lafayette, which is not that far from Lake Miccosukee, I believe he was more taken with the

Apalachicola River area," she replied. "As far as Chapman goes, it could simply have been a difficult place to find. It is a very small site. Chapman got around a lot, but travel, even in the later 1800s, was not so easy. There were several rivers and swampy lowlands to cross."

It would be 1924 before the unusual specimen with its dainty flowers and edible fruit was found and named.

On a pleasant day in early February 2020, Donna and I set out from Tallahassee toward Lake Miccosukee. We left the city via Miccosukee Road, which runs past the Croom homestead (now the sixteen-acre Goodwood Museum and Gardens) and continues northeast alongside the Miccosukee Canopy Road Greenway. The 503-acre corridor offers peaceful forest and pasturelands that were set aside by the Trust for Public Land for use by hikers, bicyclists, and equestrians on a series of looping trails. Interstate 10 crosses Miccosukee Road at the far end of the Greenway. From there we continued a short distance before turning south toward Florida Highway 90, which runs along the southernmost end of Lake Miccosukee on its way to the small town of Monticello, the seat of Jefferson County.

Highway 90 is a two-lane road that bisects large, gated estates fancifully landscaped with palm trees and pecan orchards. Interspersed among these spreads were low-lying swampy wetlands and small, 1950s-style houses and trailers, their yards strewn with abandoned cars and equipment. Here old Florida met new Florida, native and invader, but all neighbors, regardless of origin or income, shared a view of beneficent greening—moss-draped live oaks leafing out and Carolina jasmine blooming atop fences.

Every few miles, we began noticing signs with three words—NO TOLL ROAD—circled in red with a slash through the middle. Donna looked it up on her phone. The message referred to the state's plan to build the "Suncoast Connector," a limited-access toll road running from this area to Citrus County, in the middle of the state. Opponents argued that the corridor would fragment wildlife habitat, alter the flow of rivers all the way to the Gulf, and destroy ecosystem health, along with reducing local business in bypassed towns such as Monticello. Supporters, including the governor, contended that the road would encourage growth in rural areas with the potential for water, sewer, and other infrastructure development. The planned route was still subject to amendments, but legislation had already passed that would start construction in late 2022, ending in 2030.

Donna was reading this contemporary construction plan out loud to me as we passed by the entrance to the park where Florida's largest Native

American ceremonial earthwork mound, perhaps 1,800 years old, is preserved and interpreted. Highway 90, I had read, was once a path used by many tribes that later became part of the Old Spanish Trail. The ceremonial mound, while not a Miccosukee tribal monument, was located due south of what had been for several decades the center of life of the Miccosukee people (or Mikasukee, as the name of their language is spelled). The tribe, once part of the Creek Nation, originated in north Georgia. In the early eighteenth century, the Miccosukee were displaced by European settlers. They migrated to northern Florida in what is now the town of Miccosukee, on the northwestern side of the lake.

Once in Florida, the Miccosukee joined the Seminole Nation and fought in the Seminole wars against the European invaders. Like so many Native people in the South, the Miccosukee were forced by Andrew Jackson to resettle west of the Mississippi River in 1842. The tribal members who managed to escape Jackson's forced march took refuge in the Everglades, much farther south. Today's tribe of some four hundred members was recognized by the federal government in 1962 and occupies several small reservations in southern Florida near Everglades National Park, Fort Lauderdale, and Miami. The tribe has a fully functioning government and draws revenues from a resort and gaming hotel in Miami-Dade County.

Botanists use the term "relict" to describe plant populations that were once widespread. "Relict" also seemed a fitting term for the diminished tribe that shares the name Miccosukee with the native gooseberry.

On a map, Lake Miccosukee looks like the shadow of a bird: wings wide, tail clipped, flying northeast only a few miles below the Georgia state line. Miccosukee is a prairie lake: it is shallow and empties naturally during dry spells, which help to clean the lake bottom and renew plant life. "Fire also probably played an important role in scouring the organic matter from the lake bottom during droughts," said Florida biologist Todd Engstrom, though the use of fire in lake management does not occur in this region as it once did because of the proximity of major highways, where smoke could be a hazard to drivers.

Lake Miccosukee is popular with waterfowl and gator hunters. As part of the Florida Aquifer, water from the lake in 2019 was tracked flowing underground from an underground swallet or sinkhole west to the St. Marks River, near Natural Bridge, according to Michael Hill, a local fishery biologist. Hill noted that most of the lake's water loss occurs through evapotranspiration, not through sinkholes.

In 1954, an earthen dam with a spillway was built on the northwestern shore of the lake to control the water level for sportsmen, a project that was effective for a time. Restricting the natural drainage, however, causes an accumulation of decomposing plant matter on the bottom of the lake from aquatic plants that can choke out fish. The water did seem quite shallow when we got a good look at it across a wider section of the lake.

Because the populations of Florida's Miccosukee gooseberry are on private land, we did not expect to see them up close, but I had already seen the shrub before going to Florida. The botanical garden specimens at both Duke University and the University of North Carolina at Chapel Hill seemed to be flourishing, but they had not spread out in their more northern environment in these North Carolina gardens.

We turned off Highway 90 to drive around the periphery of the private property where a portion of the gooseberry natives are still growing. Some six thousand fastidiously groomed acres of longleaf pine and wetlands are dedicated to gamebird hunting and freshwater fishing. The property belongs to the family-owned outdoor outfitter Orvis Company, the oldest mail-order retailer in the United States. The land is open to visitors only when Orvis sponsors an annual shooting school for a small number of guests interested in the finer points of dove, duck, and quail hunting. The estate also holds a kennel and dog-training facility.

Farther north, on the lake's eastern shore, we came to a historic boat launch site that locals still use and that sits near another population of gooseberries. We followed a poorly maintained dirt road, winding through thick shrubs and bottomland. We parked at the top of a bluff that descended to a small clearing where the water lapped among cypress knees. I suspected the gooseberry plants were close by, but also had a feeling that there might be alligators we didn't want to meet among the spongey thickets of maiden cane on either side of the clearing.

Great gray garlands of Spanish moss swayed in the breeze like dreadlocks. Christmas berry bushes (*Ardisia crenata*), full of bright red fruit, were growing along the edge of the clearing. Later I would learn that this variety of *ardisia*, an Asian ornamental, was first offered at a commercial nursery in Florida in 1896. By the 1960s it started appearing in natural areas of the state. Birds spread the seeds far and wide, and according to the Florida Museum website, "once it has taken root, it is hard to get rid of." It was declared an invasive species in 1995 and then upgraded to a noxious weed, making it illegal to sell or propagate in the state. Has its presence had an effect on the gooseberry here? I don't know, but it helps tell the story of how so many new plants have been introduced in this semitropical climate because of their ability to grow especially well here.

We looked across the lake as the dwarf palmetto chattered in the wind. A few wildflowers were barely emerging among the beer cans, bottles, and remnant campfire cinders left by visitors. I tried to imagine what this spot on Lake Miccosukee looked like in 1924 when the gooseberry was discovered—probably somewhat similar to what we saw on that February day ninety-six years later, except for the litter.

It was a Monday afternoon in February 1924 when the itinerant botanist Roland McMillan Harper and his acquaintance, Dr. Herman Kurz, assistant professor of botany at the Florida State College for Women (now Florida State University), decided to take a ride in Kurz's automobile to visit the northeastern side of Lake Miccosukee, where neither botanist had ever been.

Harper, who was under contract with the Florida State Geological Survey, was an eccentric who, when not botanizing, spent most of his time clipping articles from newspapers and journals that he hoarded in an elaborate filing system organized by topics of personal interest. He also compulsively collected pamphlets and made all manner of lists. He was known for taking pen to paper using the window of a car or train as his desk, and he took fastidious notes on his botanical sightings.

Harper had developed his botanical eye while riding trains. He was known to disembark at one station and follow the rails back to where he had just come, gathering specimens along the way. Though he was born in Massachusetts, his family moved to Dalton, in the north Georgia mountains, and later to Americus, in the southwestern part of the state. Harper entered the University of Georgia at the age of sixteen. Despite an almost pathological fear of snakes, he was soon drawn to the swamps of the wild South. In graduate school, he conducted one of his earliest expeditions in Georgia's Okefenokee Swamp. Later he followed part of Bartram's Trail with his younger brother Francis, also a botanist, through Georgia and Alabama.

In 1913, Harper wrote with fervor about the beauty of Cahaba lilies (discussed in the next chapter), which he had photographed on Alabama's Black Warrior River at Squaw Shoals. He later registered his horror as the population was destroyed by the construction of a lock on the river that year. He was also the first botanist of his era to understand and advocate the importance of fire to the South's longleaf forests, a position for which he was mostly dismissed by his peers during his lifetime. *Harperocallis flava*, or Harper's beauty, a small, grasslike perennial, was one of his most delicate discoveries. The species—found mostly in the Apalachicola National Forest—produces a solitary yellow flower and has been on the federal endangered list since 1979.

Harper was temperamental and given to some controversial ideas, including a disdain for ladies with bobbed hair. He also developed a passing interest in eugenics. He disliked driving a car so much that he turned down an offer to work for Thomas Edison in the search for a Florida plant that might serve as a rubber substitute. John K. Small took on the project with Edison instead, as mentioned in chapter 1.

Harper did have a lighthearted side, however. He drew cartoons of himself as "The Maniac," which may have been an accurate depiction of the mania side of the bipolar disorder he likely suffered. He also made elaborate miniature houses out of pasteboard to amuse children, including the mother of his eventual biographer, Elizabeth Findley Shores, who discovered the artifacts in her mother's attic and became curious about their maker. In her biography Shores speculates on the mental illness that apparently ran in the Harper family.

"When you meet Roland Harper," Small once wrote in a letter to a friend, "do not let his off-hand outspoken attitude toward theory, a proposition, or a person surprise you. Harper is very able, but he has the unfortunate habit of generally 'rubbing the fur the wrong way' and thus irritating people generally and making many enemies."

Harper eschewed marriage until the age of sixty-five, though someone was always trying to set him up with women. Instead, he wore out pair after pair of shoes botanizing on foot all over the Southeast. He would be one of the last botanical explorers to visit the as-yet-unspoiled pine barrens of the region.

Writing in the journal *Torreya*, about the day of the discovery of the Miccosukee gooseberry, Harper explained that he was pleased to be able to explore the region with Herman Kurz, because the professor brought an automobile to the task: "My observations in the region designated as the Tallahassee red hills were practically confined to what I could see in walking out from Tallahassee and back in the same day, a radius of ten or twelve miles," he wrote. Harper would soon regret having published commentary a decade earlier about "the remarkable dearth of rare plants in this region, which has no counterpart anywhere else in the world, and ought presumably therefore to have at least a few endemic plants." He simply had not observed the bounty of unusual plants in the area when he wrote the article.

Kurz and Harper entered the woods near the lakeshore and began taking inventory of "the luxuriance of the vegetation and the predominance of deciduous trees. If it were not for the presence of the Spanish moss," Harper wrote, "anyone entering this forest in twilight could easily imagine himself to be somewhere in Ohio or Kentucky." Kurz had brought along an augur to take soil samples, and the two men continued hiking along the wooded bluff, with Harper taking notes on the variety of trees, shrubs, and herbs. "I was surprised to see a few specimens of a currant or gooseberry," Harper wrote.

In 1895 Small had documented the southernmost US gooseberry known in botanical circles at the time. *Ribes curvatum*, also known as the granite gooseberry, lived on the giant granite outcropping of Stone Mountain, near Atlanta, and could also be found "in the mountains of Alabama over 200 miles away and in a perceptibly cooler climate." On this gooseberry near the lake, Harper wrote, "the leaves were then about half grown, and we guessed that it would be in bloom about two weeks later."

The two botanists agreed to come back to this unusual spot. As luck would have it, Harper soon learned that the foremost American authority on blueberries and currants was coming to Tallahassee at the end of the month.

Frederick Vernon Coville was the chief botanist for the US Department of Agriculture, founder of the National Arboretum, and honorary curator of the National Herbarium in Washington, DC. At the time of his visit to Tallahassee, he was also chairman of the National Geographic Society's research committee, helping evaluate areas of the earth that the Society would deem to explore. Coville is perhaps best known as the first to study and understand the special needs of blueberries—the most favorable soil, climate, and pollinators. His research had made it possible for blueberries to become a valuable commercial crop in the Northeast a decade before.

When Harper and Kurz took Coville to Lake Miccosukee on 2 March, the gooseberries were in bloom. Harper wrote: "Almost immediately [Coville] pronounced it a new species. (He described it as soon as possible thereafter without even waiting to learn the color and taste of the ripe fruit which was not available until June.)"

Kurz and Coville visited the gooseberry population again at the end of March. Then, on 22 April, Harper wrote that he "conducted Dr. John K. Small and his party to the spot and found the fruit about two-thirds grown." Small apparently could not miss out on such a find. Florida had become his primary research territory, and though he mostly focused on the Everglades and its tropical specimens to the south, he made the trip to the

Panhandle and photographed the gooseberry using a black backdrop to accentuate the new plant's features.

Coville returned once more in June to collect the ripened fruit. He pronounced it sweet and juicy. Harper explained in a footnote to his *Torreya* report on these excursions that Herman Kurz's wife had earlier gathered some of the gooseberries, "and the next morning, though not quite ripe, made from them some jelly, which in both color and taste was very similar to apple jelly."

An announcement of the new fruit was published in *Science News*— a biweekly founded only two years earlier, in 1922, and issued continuously ever since. The news release was less about science and more like a product promotion for a new delicacy coming soon to American tables.

> The elevation of the humble gooseberry from its present position as a minor fruit, good only for pies and preserves, to the dignity of a table fruit on par with cherries, plums and grapes, is one of the possible results from the discovery of a new species in the woods of northern Florida, a region where gooseberries have never before been known.
>
> Gooseberries are not properly appreciated in this country is the opinion of Dr. Frederick V. Coville, botanist in charge of the Office of Economic and Systematic Botany, Bureau of Plant Industry, whose description gives the new species to science. Abroad, the fine European varieties are eaten ripe as table fruit. But these choice varieties are subject to disease in this country, and all the hardy native species known until the present time have berries too small and poorly flavored to be of much value.

The announcement goes on to explain why the species held such excitement for Coville. Blister rust, a disease affecting white pines, posed a grave threat to the prosperous timber industry in the Northeast. Because gooseberries harbored the rust and transmitted it, a federal ban had been issued in 1911 that made it illegal to grow all *Ribes* species, including currants. But as Coville realized, these gooseberries were growing in a much different climate. The *Science News* release continued:

> Into this rather unpromising setting the new Florida species comes almost like a horticultural fairy godmother. It is probably the biggest wild gooseberry ever discovered, the fruits reaching seven eighths of an inch in diameter. It is native in northern Florida, far south of the present centers of gooseberry culture, and what is even more important, far south of the white pine region. Therefore, it can be

cultivated without regard to the white pine blister rust, and there will be the added advantage of having berries ready for market much earlier than at present.

Science News admitted that there was "one notable drawback. Each berry is covered with so many long, sharp spines that it suggests a little porcupine. It is hoped that these can be eliminated in the breeding experiments now under way at the Department of Agriculture."

Todd Engstrom, a conservation biologist who has studied the Florida populations over the past decade, chuckled when he shared the nearly century-old press release with me. "I've eaten the berries," he said, "and while the spines are soft, there are these glands on the tips of each spine, that exude an unknown substance." Though we were talking by phone, I could imagine Todd's face twisting up. "Nobody knows what it is in the glands," he said.

By the time the Miccosukee gooseberry was found in the mid-1920s, Florida's tropical habitats had become a testing ground for new species that were being imported, sometimes with abandon, for the promise of new crops and expanding American appetites. Some of these introduced plants would come to pose a challenge to many native species in the state and beyond. The US Department of Agriculture had a singular impact, bringing to Florida dozens and dozens of foreign varieties of now familiar food crops—South American potatoes and new varieties of corn, for example—and exotics from around the world, including mangoes, avocados, kale, papayas, pistachios, nectarines, quinoa, cashews, dates, lemons, and nectarines—which were and are a welcome addition to our cuisine.

David Fairchild (1869–1954), an adventuresome Gilded Age botanist who worked with Coville on the National Arboretum Committee, used Florida as his primary proving ground. Fairchild singlehandedly brought in the whole list of edible species noted above to study their agricultural potential in the United States, in total more than two hundred thousand exotic plants. He test-cultivated seeds and cuttings of many of these food crops, mostly in southern Florida. His work also had the unintended consequences of introducing new pests, plant diseases, and invasive species to the nation.

Fairchild, whose career is documented in an excellent contemporary biography (*The Food Explorer*, by Daniel Stone), also worked in Hawaii with tropical species and helped to launch the commercial cultivation of exotics in the islands. "Plant Immigrants" was the title of the printed bulletin that Fairchild issued from 1908 to 1924. The newsletter, read by plant experimenters all over the country, was so named at the suggestion of Fairchild's father-in-law, Alexander Graham Bell. Partly because of Fairchild's penchant for importing invasive foreign species, Florida has the dubious distinction of being among the top states in the nation for its number of endangered plant species.

The other major long-term threat to native Florida species was habitat destruction. Harper, Small, and Fairchild shared a mutual despair as they watched the rapid commercial development of the Sunshine State, disrupting the habitats of species like the Miccosukee gooseberry that botanists were only then discovering and documenting. Small, in particular, was appalled at plant collectors who came into the Everglades to steal rare orchids and then torched the hammocks where they found them to make their stolen plants more valuable. He railed against state-funded drainage projects that converted swamplands to farmland and housing developments. Canals, he noted, were dug to make more "waterfront property" and at the same time turn wetlands into dry real estate.

From Eden to Sahara: Florida's Tragedy was Small's impassioned record of the native plant destruction and "vandalism" he witnessed over two months in the spring of 1922, two years before his initial encounter with the Miccosukee gooseberry. In the book, Small grieves the loss of Florida's Native peoples as well: "The destruction of aboriginal village sites, kitchen-middens, burial mounds, and ceremonial structures is progressing without any attempt at a scientific study and interpretation, not to mention preservation." First published in 1929, the book was brought back into print in 2004 and is in use as a reference point for conservationists who are working to preserve or restore Florida's native ecology where possible.

Apparently, nothing came of any breeding experiments that might have been conducted on the Miccosukee gooseberry after its discovery. No commercial viability for the fruit was pursued.

Meanwhile, in South Carolina, invasive privet, Japanese honeysuckle, rooting feral hogs, and browsing deer have challenged the Miccosukee gooseberry populations in the southwest corner of the state near the border with Georgia. April Punsalan, a botanist with the US Fish and Wildlife Service in Charleston, South Carolina, oversees the state's twenty-two federally endangered and threatened plants. Vivian Negrón-Ortiz, in Florida, is responsible for conducting the federal five-year review for the Miccosukee gooseberry, and April is the state lead for the plant in South Carolina, working closely with her Florida counterpart.

Unlike the Florida gooseberries, South Carolina's population of the species is permanently protected on public land. But the presence of various invasive species near the plants in the Stevens Creek Heritage Preserve and elsewhere in the Sumter National Forest presents a daunting challenge. The US Forest Service consulted with the US Fish and Wildlife Service to determine whether applying herbicide on the privet, in particular, would have a negative impact on the plant species. April told me that some herbicides might translocate from the privet roots to the gooseberry if applied in close proximity.

Privet is a common challenge to native ecosystems in the wild South, and most southern gardeners have sooner or later had to contend with this stubborn plant. My first encounter was with an expansive shrub that grew alongside the first house I ever bought in an old Durham, North Carolina, neighborhood. I engaged a friend with a landscaping business to get rid of it. Rachel arrived in a vintage pickup truck and proceeded with confidence to prune the bush back dramatically. After clearing away the forlorn limbs with their tightly clustered evergreen leaves, she attached a hefty chain around the trunk, which she then affixed to the bumper of the pickup. After she gunned the truck's engine several times with sharper and sharper force, kicking up most of the gravel in my short driveway, the chain broke. I poisoned the privet in the end, though I'm not sure I could bring myself to do that now.

According to a US Department of Agriculture (USDA) technical report, Chinese privet (*Ligustrum sinense*) was introduced in the United States from China and Europe in the early to mid-1800s and thrives from Maryland to Texas. The dense thickets tend to grow best in damp areas: low woods, bottomlands, or beside streams. It is shade tolerant and can spread as an entangled menace along fencerows. It can dominate native plants on the forest floor, even preventing the regeneration of pines and hardwoods. Privet seeds are dispersed by birds and other animals, though root runners can also expand the thickets.

April explained that privet has been shown to change the plant-soil microbiome. More happens underground than meets the eye. "Privet can kill the mycorrhizae," April said.

I had to look that up. The term "mycorrhiza" refers to the symbiotic relationships among fungi, plant roots, and the soil. The fungi in the ground add to a plant's ability to gather nutrients and water from the soil through the fungus. In exchange, the plant feeds the fungus with sugars it produces during photosynthesis.

"Privet can create a positive feedback loop in the soil," April explained, "and the seeds of other plants can't germinate when this symbiotic relationship is not functioning properly."

According to the most recent five-year report on the status of the endangered gooseberry, the South Carolina Department of Natural Resources brought in staff and volunteers to manually remove privet and honeysuckle, creating a great improvement in the publicly accessible Steven's Creek Natural Heritage Area. This labor-intensive removal by hand cleared some 70 percent of the two plant invasives in the afflicted area, but the process will require ongoing management. Privet stumps were also painted with the herbicide glyphosate to kill the stubborn roots.

Protecting the South Carolina populations against deer and feral hogs is also a labor-intensive process. Trapping feral hogs and fencing off patches of the gooseberry from deer and hogs in South Carolina has been recommended in the latest report from the US Fish and Wildlife Service by lead recovery botanist Negrón-Ortiz.

Come early spring, the 1.8-mile Stevens Creek Heritage Preserve trail becomes a favorite hiking spot for dedicated South Carolina naturalists. Here the Miccosukee gooseberry can be seen from the footpath. The soil there—with its high pH and unusual concentration of calcium in the rocky outcroppings—is similar to other extremely biodiverse hotspots discussed in these pages: along the Cumberland Plateau in Tennessee; at the Keel Mountain Preserve in Alabama; and in the native habitat of the Florida Torreya tree. All of these hotspots offer the chance to experience rare or unusual wildflowers normally seen in colder climates. Dr. Patrick McMillan—former director of the South Carolina Botanical Garden and host of *Expeditions with Patrick McMillan* on public television—has visited the preserve more than thirty times and has written that every time he hikes the trail, he sees something new. Seasoned wildflower buffs know to plan their hikes for mid-March to early April to see the gooseberry blooming.

Protecting the gooseberry in Florida has required intensive, if inconclusive, research. Among the most distinguished contemporary conservationists working to protect Florida's remaining natural assets is Wilson Baker, now in his eighties and credited with discovering several previously unknown plant and animal species in the state. Baker was the first full-time biologist hired at the Tall Timbers Research Station, north of Tallahassee, in the 1960s.

Tall Timbers is known for its work with prescribed fire to manage for species biodiversity, selective timber harvesting, and habitat protection for the bobwhite quail, popular with hunters. Baker's contributions at Tall Timbers have focused on the biodiversity of the longleaf pine ecosystem.

Tall Timbers works with many partners in the larger Red Hills region, which encompasses two counties in Georgia and three in Florida, including Leon and Jefferson Counties—the remaining gooseberry habitat. The Red Hills, which run from Tallahassee to Thomasville, Georgia, have the largest contiguous acreage of native longleaf pine left on privately owned land in the country, including 100,000 acres protected by conservation easement. These 436,000 acres of rolling red clay hills have been designated by The Nature Conservancy as one of America's "Last Great Places" and named among the top ten scenic corridors in the United States by Scenic America. The Red Hills are still yielding their botanical secrets to researchers from Florida State University and Tall Timbers.

By the beginning of the Civil War much of this land—including Hardy Croom's Goodwood plantation and dozens more agricultural ventures—had been disturbed and put in the service of growing cotton. It would take time for the landscape to recover, a process underway by the time Roland Harper came hiking along. Today the region has the largest concentration of undeveloped former plantation lands in the United States and is still in recovery.

In 2010, Negrón-Ortiz called on Tall Timbers's Wilson Baker to investigate what was causing a decline in one population of gooseberries at the property in Jefferson County. Baker in turn subcontracted the biologist Todd Engstrom. The two scientists, once mentor and apprentice and by then good friends, had worked together on the reintroduction and protection of the red-cockaded woodpecker and its habitat in the Red Hills.

In our interview, Todd said he had never really worked with plants before the gooseberry assignment but discovered with pleasure that he had a good eye for the rare plant. He began his work for Fish and Wildlife by conducting a systematic survey, individually mapping each gooseberry plant along the lakeshore and the two subpopulations on hunting land. He plugged the plants' positions into a geolocator for future monitoring.

"I observed that the gooseberry can become abundant in a light gap caused by a treefall," Todd said. "The stems take root where they come into contact with the soil." But he also found many plants in various stages of being battered by falling limbs, encroached on by moss, and sitting in the shadows of large trees blocking out the necessary sunlight. In addition, an extremely dense, midlayer growth of cherry laurel was shading out the gooseberry in places. Todd and a graduate student, Tom Radzio, found an extreme range in the numbers of fruit produced on individual plants. Very few plants were truly productive.

An earlier study by Canadian scientist P. M. Catling had identified several kinds of pollinators working on the gooseberries when they bloomed, but the relatively small numbers of the plants and the locations of the plants in their isolated patches might account for the poor fruit production Catling suggested.

I asked Todd whether pollution could be a factor. "Most of the plants are not close enough to agricultural fields to be damaged by chemicals," he replied. Another study found that occasional saltwater intrusion had occurred at Lake Miccosukee with hurricanes over the years, but that was not a likely factor in the decline either.

Based on genetics, the authors of one study, including Vivian Negrón-Ortiz, suspected the declining population might be due to inbreeding, or put another way, to the lack of genetic diversity because the plants are now so few and so isolated.

"How many true genetic individuals are there among these gooseberries or are they just sending out shoots from the original plant? That was the question," Todd said. Would introducing genetic variation reinvigorate the Florida plants and create new seedlings? Todd waited until the right moment and headed north to the Stevens Creek Heritage Preserve in South Carolina to collect pollen from the blooming Miccosukee gooseberries there. He hoped to offer an alternative to mating such closely related individual plants in Florida by using the genetically different pollen from South Carolina. "I rushed back to Florida with the pollen, but I was not successful in my pollination efforts," Todd said.

Finally, Todd found a very productive plant with sixty-five gooseberries on it. The shrub was in a gap where a tree had fallen, and the trunk was right up against the unharmed gooseberry. They wondered, Why didn't this plant with all the fruit (and seeds therein) spread to make more plants? Why didn't the range of the plants expand into the area where suitable habitat existed both north and south of the present stand? The single superproductive plant with plenty of sun to develop its fruits continued to make abundant gooseberries over several years. Then the botanists realized that its fruit was being eaten by something. Was it a bird or was it perhaps an insect pest, depositing larvae that then ate their way through the berries?

No evidence of insects turned up, so Todd positioned an infrared camera at the base of the fruited gooseberry. The camera took a picture every second.

By training the camera on the plants at night, Todd did identify a predator that was eating the fruit and chewing up the seeds so thoroughly that they could not possibly germinate later. "We were lucky," Todd said. "At two in the morning, we could see the stem starting to shake. And then here comes a cotton mouse enjoying the fruit."

Cotton mice make their homes on the forest floor in upland areas of the Southeast. They earned the name by their habit of using raw cotton as a nesting material. They eat seeds and insects. Cotton mice are also among the species who prefer underground refuges, such as those provided by gopher tortoises, by which they can even survive natural or prescribed fires in the longleaf forest. The predation by cotton mice that chewed up the seeds too thoroughly for them to sprout, however, did not seem as significant as the poor survival rates of seedlings among the southernmost patch of plants in Florida.

Research continues on the many threats to the Miccosukee gooseberry, but it seems that so far none of those has struck a fatal blow. Negrón-Ortiz's most recent study of the declining subpopulation in Florida "suggests that there is no persistent soil seed bank that can be relied on to maintain populations," though she explains that other members of the *Ribes* species have survived many years of dormancy in the ground and in storage. Sierra gooseberry seeds, she writes, remained viable in California after being stored for forty years in bottles.

Though one of Florida's subpopulations is declining in numbers and the other is somewhat improving, the overall scarcity of plants in Florida is cause for alarm. Further study is needed to determine the role of precipitation (or lack thereof) in the growing season and other factors that may contribute to poor seedling survival. Understanding more about the role of downed trees in improving gooseberry production is needed, as is research to determine the impact of the overstory in shading the plants.

Bottom line: the limited number of plants remaining in Florida creates tremendous pressure on species' survival and leaves little room for the unpredictable fluctuations in environmental factors such as temperatures and rainfall in this tiny sliver of land in Florida where the gooseberry is holding on. The primary natural threat to the plant here remains a mystery, though, as Negrón-Ortiz put it in her five-year report, because the gooseberry lives exclusively on private property in Florida, the human threat of habitat alteration or destruction is most critical. "There is no guarantee that the properties will not be developed for home-sites, agriculture, logging of associated hardwoods, recreational facilities or other purposes in the future, although the owners have not given any indication that they intend to do so," she writes.

Given all these factors, permanently maintaining plants and seed materials from all the known populations of the Miccosukee gooseberry in nurseries and botanical gardens has become essential. These collections may be tapped for reintroduction of the plant if the populations continue to decline. Long-term seed bank storage and other measures, such as building and maintaining partnerships with private landowners in the native range of the plant, may be increasingly necessary.

At the end of our Florida investigations, we drove from Lake Miccosukee southwest to meet with author Gail Fishman, who was working as a ranger at the St. Marks National Wildlife Refuge. Along the way, we sped by many flat acres of longleaf pine barrens and miraculously came upon wild, short-stemmed Atamasco lilies, also called rain lilies (*Zephyranthes atamasco*), blooming along the roadside at odd intervals. Triggered by a recent rainfall, they were a visual blessing on our journey.

We stopped for oyster stew at a rundown restaurant next to the St. Marks River bridge, where the view of the water looked pretty much the same as in a photograph that John K. Small had taken in 1924. As we entered the bar, a patron, apparently a regular, asked if we were "lighthouse people." "You look like lighthouse people." We laughed.

St. Marks Wildlife Refuge is a popular destination for the sighting of magnificent herons, turtles, ducks, alligators, and cranes along the drive to the historic lighthouse where the river meets the Gulf. After lunch and before we made our pilgrimage to the lighthouse, however, we met Gail at her office in the St. Marks Visitor Center. She took us out on the deck overlooking the wetlands and pointed out a female alligator, one of a resident pair that the volunteers in the gift shop enjoyed watching from a distance. In Gail's office, we sat down to get acquainted. I thanked her for her absorbing and eloquent book. A Florida native in her seventies, she was dressed in the uniform of the US Fish and Wildlife Service: brown pants and a khaki shirt, which set off her thick white hair.

During her career as a freelance writer, Gail worked for the Florida Defenders of the Environment, The Nature Conservancy, and the National Audubon Society. She was also the only person I met who had visited E. E. Callaway's Garden of Eden. Her father took her to see the Torreya trees as a child. Apparently he saw it as an early rite of passage for a would-be

Florida native Gail Fishman, author of Journey through Paradise: Pioneering Naturalists of the Southeast, *has spent the last years of her career as a ranger for the US Fish and Wildlife Service at the St. Marks Wildlife Refuge in the Panhandle.*

Florida natural historian. The trip made an impression. Gail has been trying most of her life to help stem the tide of environmental destruction in her home state. With some emotion she explained that many years later, when she was diagnosed with uterine cancer, she was prescribed Taxol, the drug based on the compound from the bark of the Torreya's rare relative, the Florida yew tree. "It saved my life," she said.

Gail chose to write about the lives of the early botanists to give something back to her home state and the natural assets that have forged her identity. "I wrote my book because people needed to know who these men were." She sat at the feet of the naturalist Angus Gholson when she worked at the Apalachicola Bluffs and Ravines Preserve. "Angus tried to teach me the Latin names of all the plants that Croom, Chapman, Small, Harper, the two Bartrams, and the two Michaux found," she said.

Writing such an elegiac book was distressing, however, because it required that Gail go back to places where she had not been in years and where she saw the drastic changes wrought on the state in her lifetime.

Gail still grieves the loss of the Florida she knew as a child. Her father preferred to travel off the main highways for the family excursions that he loved to plan and execute. For these trips, her mother packed picnics and cold Coca-Cola in six-ounce bottles, and they would set out before dawn to drive from one end of the peninsula to the other. "To my young eyes," she wrote in *Journey through Paradise*, "Florida seemed a vast wilderness in the middle 1950s, and in many locales it was."

To write her book, especially the chapter about Small, Gail journeyed to the southernmost parts of the state. She concluded, "These days, traveling much farther south than Gainesville in my home state leaves me drained, disappointed, frustrated, saddened, and distraught—any emotion but happy."

Over the years, Gail has watched as her contemporaries and now younger native Floridians sell off property they inherited from their parents. "They don't want it; maybe they have moved elsewhere," she said. "These are land sales that lead to more development." Now Gail takes comfort in the daily rhythms of the wildlife preserved at St. Marks Refuge.

"I'm skeptical of the pigeon and the parakeet being seed spreaders for the gooseberry," she said, as our conversation turned toward endangered species. "Another option is perhaps that Native people traded the plants and seeds, and that is how the species moved from one state to another."

Gail might be on to something. The Stevens Creek Heritage Preserve in McCormick County, South Carolina, is very near the route that William Bartram followed when he crossed the Savannah River in 1775 and headed south into Georgia. Bartram documented coming across a heavily traveled trade route that Indigenous people used. Bartram's historic path in that area—much of which is now underwater—is marked for visitors starting in a parking lot on the Clarks Hill Reservoir. The site is only seven and a half miles by car to the Stevens Creek Heritage Preserve. Perhaps the Stevens Creek area, though steeply sloped in places and peppered with precipitous marble and calcium rock outcroppings, was a place of refuge and berry picking for the earliest peoples in the region. The Miccosukee people, as far as is known, started out in Georgia and then headed south to Florida when European encroachment forced them out. Todd Engstrom did say that in his experience *Ribes echinellum* has been easy to grow from seed. So perhaps the Creeks or Cherokee or Miccosukee obtained seeds in South Carolina and carried them to Florida.

"It's a complicated question, and the answer might never be found," Gail said. "Trade between tribes went on for a lot of years before Europeans ever got here. Speculation keeps us thinking, though."

In a December 1920 report that followed one of Small's many botanizing trips to Florida, he called the state "Land of the Question Mark."

A glance at the map of the continental United States will show that Florida—the most southern State of the Union—suggests the shape of an interrogation mark. There is a fitness in this. The geographical position of the State, particularly of that great tongue of land thrust southward hundreds of miles into tepid seas, implies a plant-covering unique in North America, and thus makes the question mark—so far as botanists are concerned—both a challenge and an invitation. Not one, but myriad questions, moreover, are suggested by that note of interrogation.

And questions, in many cases, are all we have left about the species that might have gone completely undocumented in the twentieth-century transformation of Florida to the current man-made "paradise" of high-rise condos, boardwalks, inland canals, golf courses, airboats, swimming pools, and family amusements. Though the Miccosukee gooseberry was given a name, it is precipitously close to joining other plants no longer present in the wild South. With that loss, should it occur, will go a raft of stories about Native Americans, settlers, botanists, and contemporary visitors.

5

Shoals Spider Lily (Cahaba Lily)

 Every year between Mother's Day and Father's Day, a rare aquatic lily (*Hymenocallis coronaria*) blooms in wide clusters on four major rivers in the South. The stems of this lily, a member of the amaryllis family, are tall, between two and four feet. Their bulbous roots hold fast in the cracks of dark rocks in shallow shoals where the river is swift. Come evening, the day's new blossoms open, elaborate and fragrant. As the air cools and a breeze picks up, they sway like flocks of preening birds atop their leathery green stems. Their perfume is as intoxicating as honeysuckle.

The flower's corona is three inches wide, with six white ribbon-petals trailing outward at equal intervals around the scalloped edges. In the center of the bloom, a yellow-green eye gazes from behind the dangling stamens, their anthers coated with an orangish pollen. It's an extravagant spectacle for most of a month. Up close, the blossoms are best seen from a kayak or canoe if you dare to navigate among the rocky islands of flowers. Each bloom is at its best for only a day, but every plant produces six to nine buds over the course of its late spring performance.

The scarcity of this perennial lily illustrates the impact of population growth along our rivers in the once-wild South. According to the principal expert on the species—Professor Larry Davenport of Samford University—there are only seventy stands of the shoals spider lily left in the world. Known as the Cahaba lily in Alabama (named for the primary river where it still blooms) and as the shoals spider lily on Georgia's Broad and Flint Rivers and on South Carolina's Catawba River, the species has been greatly reduced by increasingly dense residential and commercial development and the degraded water quality, silt accumulation, and stormwater runoff that follows. Among the seventy stands of lilies, some smaller populations still occur in tributaries and creeks connected to the major rivers.

Deep water is no friend to the lily. Hydroelectric dams that were built early in the twentieth century have wiped out whole populations on other rivers in the region where they once thrived. Increased flooding from severe weather events in recent years has also been known to put the remaining plants deep underwater at bloom time. Nevertheless, devoted pilgrims come every year to pay their respects to this natural phenomenon where it still occurs, and in so doing they have created a seasonal economic bump in some less-than-flourishing rural towns.

Fortunately for the shoals lily in South Carolina, a segment of the Catawba River along the fall line remains friendly to the species. At Landsford Canal State Park in Lancaster County, a stretch of shallows a half-mile wide on the Catawba hosts nearly thirty acres of lilies. The state park was not created because of the rare plants, however, but because the area was once part of a historically significant inland transportation corridor.

Beginning in 1820, South Carolina engineers employed local enslaved people and other laborers from the North to construct a canal route parallel to the shallow rapids at Land's Ford on the Catawba. Fortified with heavy rock and masonry, the canal's five locks dropped thirty-four feet over a two-mile stretch, providing a more efficient way for small freight and passenger barges to bypass the rapids at the site. Completed in 1823, the canal was used only fifteen years before one of the locks collapsed. A massive stone bridge over the canal remained accessible, however, as part of the Great Philadelphia Wagon Road, which ran all the way from Pennsylvania to Augusta, Georgia. A railroad line eventually supplanted the need for riverboats to carry goods on the Catawba.

Earlier, Native Americans on trading missions had crossed their namesake river at this site. To this day, the Catawba tribe occupies a reservation close by, as discussed in chapter 9. After European settlers began to dominate the region, British and American forces crossed back and forth here as they fought each other during the American Revolution. Sherman's soldiers also forded the river at the shallows during the Civil War.

The canal, the only one remaining from a larger system that once linked Charleston to the Upstate, was listed on the National Register of Historic Places in 1969. The South Carolina Department of Parks, Recreation, and Tourism manages the four-hundred-plus acres of Landsford Canal State Park, now a wooded preserve threaded with trails and campsites. The modest visitor center, housed in a log cabin dating back to the 1790s, was moved from the nearby town of Chester to its present site and offers a pleasant porch with rocking chairs overlooking the water. A path along the riverbank leads to a wooden observation platform overlooking the largest single mass of shoals spider lilies remaining on the planet. The Katawba Valley Land Trust, a nonprofit conservation organization in South Carolina, helped to expand the protected area around Landsford Canal State Park in Chester and Lancaster Counties from 200 acres to more than 1,400 acres through partnerships with private landowners and federal and state agencies.

When Donna and I arrived on a weekday at the end of May 2019, we first drove to the north end of Landsford, where a substantial parking lot and boat put-in provide for visitors who come for river recreation in warmer weather and to see the lilies up close in May. Boaters can run the rocky rapids for more than a mile downriver, where a takeout allows portage back to the parking lot on a wooded trail. Venturesome boaters and trail hikers can continue another three miles south of the takeout to Highway 9 at Catawba Lake, just west of the town of Lancaster.

The headwaters of the Catawba River are in the Blue Ridge Mountains. Beginning in the early twentieth century, Duke Power Company, now called Duke Energy, built eleven dams along the river: seven in North Carolina and four in South Carolina. The large recreational lakes created by the dams have been developed steadily over the years to include parks, marinas, second homes, and subdivisions. To accommodate recreational paddlers on summer weekends, Duke Energy controls the release of the river water coming through Landsford from a hydroelectric dam upstream on Lake Wylie, along the North Carolina/South Carolina border. But when the lilies are blooming in early spring, rains may make the water levels too high for boaters (or for the lilies, for that matter.)

At the visitor center in the middle of the park, several small groups of guests milled about in the afternoon heat, including two women from Germany who had just come back from a hike down to the shoals to see the lilies. Giddy with the spectacle, they told us they were touring out-of-the-way destinations in search of small-town America, and soon headed off to their next one in Georgia.

Wayne Lanier, a park volunteer, was observing the comings and goings near the visitor center. He greeted us and explained that he and his wife, Judy, had come here every year since their retirement to live in the campground and to help the park ranger manage the rush of guests in May and on through the summer. Wayne, a heavyset, unshaven man with the look of someone completely happy living out of a camper, was ready to tell us more but was distracted by a flash of yellow feathers in the forest beyond us. I followed his gaze.

"Prothonotary warbler," Wayne said.

We shared the momentary thrill, realizing that this swamp bird was probably working a nest in the area. The prothonotary warbler builds its nests in the holes of dead trees. The name refers to the bright yellow robes worn by Roman Catholic clerks, or prothonotaries, who serve the pope in Rome. Because of the species' declining numbers in the Southeast, many birders are eager to add this sunny creature to their life lists. I had never seen one, but had heard about it from a naturalist friend, who often leads spring tours to the most remote parts of the Black River, in eastern North Carolina, to find these birds.

I explained to Wayne that we had come to experience the lilies for the first time. He smiled and said our timing was excellent. This day might be the very peak of the bloom time for the year. Donna and I high-fived. Wayne encouraged us to watch for the resident bald eagles that might be feeding on the river in the shallows. "The female eagle has laid eggs in the same nest every February since the mid-1990s," he said, "and the eaglets always fledge in late May. She had two babies this year, but the nest got too heavy and fell out of the tree in a windstorm a few weeks ago." (The fledglings survived.) Wayne told us that the parent eagles, with wingspans of seven feet, dive toward the river at speeds close to one hundred miles per hour and can carry away a fish as heavy as four pounds. Then he pointed us toward the trail.

The cut along the riverbank was flanked by river cane and small trees, some angled perilously toward the water, apparently leaning from washouts in past floods. Farther into the forest, we saw the mottled trunks of sycamore, tall tulip trees (*Liriodendron tulipifera* L. *Magnoliaceae*), and smaller holly. The whisper and gurgle of rushing water accompanied us all the way, only once disrupted by the louder drone of a low-flying airplane, heading toward Charlotte's Douglas Airport, only thirty miles north. Unfortunately, the river smelled of effluent and was soapy looking where it swirled into eddies. Seeing these whitish areas on the water at a distance through the trees, I thought at first they might be the lilies, but it was only a tease. The sewer smell was to be expected. We were not that far downstream from North Carolina's largest city and its suburbs, which extend into South Carolina.

In 2008, the environmental group American Rivers named the Catawba the most endangered river in the nation. Sewage spills and swimming advisories continue from time to time along the Catawba, especially near

Charlotte. Only two weeks after our visit to the park, at the height of June rains, some one hundred thousand gallons of raw sewage flowed into the river in an area of dense suburban development above Charlotte. Several residents on Mountain Lake had to be rescued from rising waters, and boaters were told to stay off the water for several days because of floating debris.

Farther down the hiking trail, a sealed but empty bottle floated by us— no message, except its very presence in the swift water. We met only a few other hikers heading toward us, including a wet couple who carried a red kayak over their heads. A happy child marched between them.

At a clearing, we stopped to watch a kingfisher dart over the water. A woodpecker hammered behind us in the woods. The squirrels who were foraging in the leaves along the bank seemed to be moving in slow motion because of the increasing afternoon heat, and later an unpleasant fish smell hit us. I wondered if we were close to the eagles' territory, with leftovers from their supper still somewhere on the rocks. As we drew closer to our destination, I thought I heard an eagle's cry, but I could not spot the source through the tree canopy.

When at last we arrived at the observation deck, we were alone. The sun, coming in and out of slow-moving clouds, threw bright spotlights on different areas of the blooming clusters, close and far, across the half-mile-wide river. It was too much to take in at once. *So this is what thirty acres of flowers looks like*, I thought. I wished for a boat. The islands of blooms seemed to extend beyond what we could see in either direction at this bend in the river. Their white faces were uncountable and miraculous. So much bright beauty to take in—bushels of long-stem bouquets hovering above the rocks, nodding in the light. The painterly clouds above were no match for the spectacle on the water.

Just three months before, Donna and I had traveled to see the annual visitation of snow geese and tundra swans that migrate from Siberia, Greenland, and other parts of the Arctic to spend winter in a remote wildlife refuge in eastern North Carolina. Thousands of birds convene every year in a long, white regatta on Pungo Lake and feed in neighboring fields of grain, their incessant honking carried on the harsh February winds. At one moment while we were watching them from afar on an observation deck much like the one where we stood on the Catawba, the birds had been startled—perhaps by an eagle, we were told—and they all rose into the air at once. It was like a sudden blizzard or a blinding ticker-tape parade over the water.

And now this scene. I realized I had witnessed two natural miracles of a

Larry Davenport, a Cahaba lily expert at Samford University in Birmingham, Alabama, stands beside a painting of the rare lilies blooming on the Cahaba River at Hargrove Shoals. Samford bought the painting, by the English-born artist Julyan Davis, to commemorate Davenport's years of research on the local species.

lifetime in the span of a few months. The exhilaration was uncontainable. Donna could not stop taking pictures. The similarity between these two manifestations of the unfettered natural world filled me with hope and humility. I felt very small.

Why are these incredible lilies—so scarce and fragile—not on the federal endangered species list? We went to Alabama to ask the expert, Larry Davenport, at Samford University.

Samford, a Baptist school, is Alabama's top-ranked private university. The gated campus is striking, with immaculate green lawns and vast beds of azaleas set in islands of pine straw. The tidy brick buildings rise alongside a tree-lined boulevard in the Homewood suburb of Birmingham.

Professor Davenport, the senior faculty member in the natural sciences, has spent his career here since earning a PhD in biology at the University of Alabama in 1985. He's accrued the deep affection of generations of students and the respect of the broader Alabama public. In May 2006, the Alabama legislature declared a single day (when the Cahaba lilies were blooming) Professor Lawrence J. Davenport Day. In 2007, he was named Alabama Professor of the Year by the Carnegie Foundation for the Advancement of Teaching.

Though he is prone to heap praise on his conservation partners in the work, Larry has been the de facto press agent for the Cahaba lily for more than three decades, and his is a steady and solemn voice on the fate of Alabama's endangered plants and the emerging effects of climate change.

He works out of Samford's stately, 96,000-square-foot science center, which houses a medicinal plant conservatory at one end of the building and an impressive Imax theater at the other end. On the office door next to his, three bumper stickers hinted at how religion and science—so often at odds in the South—find common ground at Samford. They read, in this order: "Your Soul Needs the Wild"; "If You Love the Creator, Take Care of Creation"; and "Extinction Isn't Stewardship." On a nearby bulletin board, posters for campus events were pinned up alongside copies of the latest entries from Larry's long-running column "Nature Journal," which has appeared in the popular magazine *Alabama Heritage* since 1993.

Larry soon ambled down the hall, dressed in a tartan shirt, black trousers, and black wing-tip shoes. We had agreed to meet following his early Monday morning class. On his office door, a single sticker had been posted, as if in conversation with the messages of his next-door colleagues: "Trust Me, I'm a Botanist."

Larry is an affable man with a scruffy gray beard and eyes that wrinkle at the edges behind his wire-framed glasses as he smiles. He told Donna as he let us into his office to be sure and capture his best side in her photo. He'd managed to get several cuts on his hands and a gash on one side of his bald head. He'd been traipsing through the woods over the weekend, he said, shrugging.

Larry is not a southerner, though he speaks slowly and has mastered a deadpan delivery of stories that often mask his intended irony. He was raised in sight of the Cascade Mountains and Mount Rainier, on Lake

Washington in the Pacific Northwest. His father was a forester and surveyor, and the family traveled mostly by boat and ferry when he was a child. They also owned a cabin in the woods of western Washington and spent, he said, "too many weekends there."

When Larry was fourteen, the family moved east to Chicago, and two years later to Indiana. After high school, he enrolled at Miami University in Oxford, Ohio. "I went as an I-don't-know-what major," he said. His freshman year he was required to take a science course. Since he'd already experienced "the obligatory frog pithing and pig cutting" in high school biology, Larry made a split-second decision to take botany. "I took the green course," as he put it, and that has made all the difference.

His lab instructor was a young PhD named Hardy Eshbaugh. "The first day of class we met for about five minutes. He passed out the syllabus and then said, 'Let's take a walk.' We hiked all over campus for three hours. He showed us different trees. He talked about them. We asked questions, and at the end he said we'd have a quiz in three weeks." Larry paused. "I had never seriously looked at a tree before, and I remember thinking, 'So this is a sycamore and this is a sugar maple—gosh, I could never tell them apart.' But I was fascinated by it, and I was fascinated by him, because he was cool, you know? I started projecting, 'Well, maybe I can do this, maybe I can be him.'"

Larry Davenport became one among scores of devoted students who followed Hardy Eshbaugh into the wilds of botany. Eshbaugh went on to become a distinguished researcher, director of the herbarium at Miami of Ohio, and a passionate conservationist.

During graduate school at the University of Alabama in Tuscaloosa, Larry had planned to save the world through ethnobotany: the cultural study of how people connect with and use plants. "Then I got into basic plant taxonomy, systematics, and finally specialized in wetland and aquatic plants," he said. For his dissertation, Larry selected *Hydrolea*, a handsome blue-flowering perennial that is often used in wetlands restoration. He traveled to Mexico and South America for research and polished his Spanish in the process.

Even before defending his dissertation, Larry was hired to teach at Samford, and three years later he published his dissertation. Then he started looking for another research project to occupy his botanical appetites. A campus colleague mentioned that the Fish and Wildlife Service (FWS) out of Jackson, Mississippi, was looking for a botanist to study the Cahaba lily.

"I had seen it once in grad school," Larry said. "It was fall, and the plant was in fruit and not very impressive." FWS offered him basic travel costs to study the plant across Alabama and Georgia. The lily was a candidate species for the endangered list. No one had ever studied the plant with much rigor, though William Bartram first wrote of its heady fragrance in 1793, when he saw it blooming in the Savannah River basin. For the plant to be considered for the endangered list, an FWS staff member or a private firm or university would have to pull together all the basic information about its history, its current populations, and threats to its survival.

Larry leaned over his desk and shook his head. "It was quite an adventure. I engaged in all kinds of dangerous activities that I would never do again." He had two near-drowning experiences in separate rivers and lost two cameras in the rapids during the research phase. "The plant lives in a very precarious situation," he said, "which adds to its mystique."

Larry tracked down as many populations as he could find. At the Flint River in Georgia, the owner of a paddle business shut down his shop and took Larry to see the lilies there. He documented "when it flowered, when it fruited, and whether the fruits need an overwintering period, which they don't. One of my greatest discoveries early on, and I'm very proud of this," he said, tongue halfway in cheek, "was when I gathered fruits from both Cahaba lilies and its sister species, *Hymenocallis occidentalis*—which I call the swamp lily. I had sets of Cahaba seeds and swamp lily seeds. I believed there must be some important biological difference between these sister species, which some people confuse when they see them blooming. So I put both sets of seeds in the same large beaker. One set floated while the other sank, making a dramatic demonstration. I took a picture of my big experiment on the steps of the old biology building." Larry sat back in his chair and crossed his arms, leaving us to ponder the significance.

Further study confirmed that the developing Cahaba lily seeds weighed down their stalks, which would droop into the water. Then the seeds fell off individually and sank. The seeds were then wedged into the cracks and crevices of the rocky shoals by the racing current and germinated within a week or so. "Meanwhile," Larry said, "swamp lilies have seeds that float, and when the swamp waters recede, those seeds stick like a bathtub ring around the edge of the wetlands."

Larry grinned. "It's my one lightbulb moment. We are all given one, and that's mine," he mused. The sinking seeds accounted for the lily's presence on the same rocks year after year.

"But what I didn't realize early on," Larry concluded, "was just how important this plant is to the people of central Alabama." He paused, his eyes welling. "I mean, it's sacred stuff, and people are very protective about it."

The night before we met Larry at Samford, Donna and I ventured to West Blocton, Alabama, site of the annual Cahaba Lily Festival. The town is in Bibb County, due south of Interstate 20 as it runs southwest between Birmingham and Tuscaloosa. The nearest sign of economic life seemed to be at the exit just beyond West Blocton, in Vance, Alabama, where Mercedes Benz employs 4,200 workers from Bibb and Tuscaloosa Counties in the manufacture of luxury vehicles. West Blocton, by contrast, has a population of some 1,200.

The streets of West Blocton were nearly deserted, save for a few middle school boys on a Sunday afternoon jaunt with tree limbs for walking sticks. The commercial district—including an old theater with its historic balcony once designed to segregate Black folks—was permanently shuttered. Only a shade-tree mechanic and a popular drive-in called Lemley's Tiger Hut provided any regular commerce on Main Street. At the restaurant, the eclectic international menu was served from a curbside window—burgers, fried pickles, taco salad, Polish dogs, egg rolls, and a much-touted lemon pepper chicken. Tiger Hut was closed on Sundays, however.

Dr. Lisa Buck, our Airbnb host, owns adjacent buildings on Main Street. She converted one into her residence and the other into a spacious apartment for guests. She also happens to be the caretaker of the Cahaba Lily Center down the block. The long-gone department store has been transformed into the town's lily museum and meeting hall for gatherings related to the festival.

Originally from Northport, Alabama, across the river from Tuscaloosa, Lisa earned her bachelor's and master's degrees in music education at Jackson State University. She then moved to West Blocton to direct the high school band. She played woodwinds for years, but now favors the ukulele. Her string band performs during the lily festival each year. Lisa switched midcareer from teaching music to giving online career preparedness classes for ninth graders when she completed a doctorate in instructional technology.

Lisa fell in love with West Blocton and told us she plans to spend the rest of her life there. Dressed casually in a West Blocton High hoodie, jeans, and hiking shoes, she favors the outdoors and is a local history buff. She drives a Honda Element, "the one car you can scoot out with a hose," she says, and takes it car camping every summer. Her proudest activity beyond teaching, however, is helping to run the events on Lily Day, when three to four hundred people descend on West Blocton to hear Larry Davenport speak, eat a potluck dinner, watch the crowning of Miss Cahaba Lily, and shuttle back and forth the rest of the day to the banks of the Cahaba River, four miles east of town. Lily celebrants, including older folks pushing walkers, will wade into the water to observe the lilies up close. Some bring lawn chairs and soak their feet in the shallows while admiring the flower show. When we visited, the town was preparing for the thirty-first annual Cahaba Lily Festival, scheduled for 16 May 2020.

Lisa took us to the river first thing, along the way pointing out the historic district where a booming coal-mining operation gave birth to the town. The name "Blocton" refers to a miraculous, one-ton mass of coal that was excavated there in 1884 when the town's population was around 3,600. Beehive coke ovens were dug out of the red dirt banks and are still visible in the district where a railroad spur once carried the coal to Birmingham.

Sadly, the town burned in 1927, and in 1928 the coal operation lost its biggest buyer. The crash of 1929 basically finished off West Blocton's industry. Citizens at the time began "borrowing" bricks from the coke ovens to rebuild the town, while hobos took shelter in the abandoned ovens. Later, groups of men would hold cockfighting matches in the ovens, Lisa said.

County Highway 24, lined with dense woods, descended like a roller coaster as we came toward the river. Before the bridge, Lisa slowed and pulled onto a one-lane dirt road that paralleled the water. Recent rains had gouged out deep gullies, but she told us that she had ferried so many visitors to this spot over the years, she could drive the road in her sleep. We bounced over rocks and rises and finally came around a blind curve to reach a popular beach. A mile farther down the river, Hargrove Shoals has the Cahaba's largest span of blooms. Despite the claims of Catawba River naturalists that South Carolina has the largest single collection of plants, Larry Davenport has declared that Hargrove Shoals is home to the largest population of the species in the world. To reach Hargrove Shoals, Lisa explained, you have to ford Caffey Creek where it merges into the Cahaba—which was running fast and deep that day—and then keep walking (or paddling) downstream.

In total, about a quarter of the entire lily species lives on the Cahaba. It is one of the few rivers in the state that does not have any high dams along its 190-mile course. Other Alabama rivers—the Black Warrior and the Coosa—have lost most of their lilies to damming and deep water. The Coosa also lost more than half of its sixty species of snails, as we later learned from Beth Stewart of the Cahaba River Society (CRS). "Everything that lives in a river needs a way to get upstream," as Beth put it, "but dams or impoundments make that impossible. Some fish serve as a parasitic host for our rare mussels that deposit eggs on them to ride upstream for propagation. Even bugs fly upstream to repopulate the headwaters of our Alabama rivers," she said.

The Cahaba was high and muddy on that February afternoon, but we could tell the rocky shoals were not far below the rippling water. A couple of young boys and their father, a photographer, were hunting for shells and reptiles along the sandy beach above Caffey Creek. The Cahaba, we learned, is home to two species of snails found nowhere else on earth, and the river has retained all of its original complement of snail species, which no other similar Alabama river system can claim. The beach presented quite a colorful collection of small mussel and snail shells at the water's edge. The boys were having fun digging moats and stuffing shells and rocks in their pockets.

"This is a Mayberry situation," Lisa said, noting how the river had become a popular year-round site for families to bring their children and experience the outdoors in this, the smallest county in Alabama. We could see people high above on the bluffs across the river, looking down from observation decks. One of the proudest outcomes of West Blocton's Lily Festival, both Lisa and Larry had told us, was the establishment of the Cahaba River National Wildlife Refuge: some 3,690 acres legally protected since 2002 and managed by the FWS. The platforms that we were looking at high above the river were connected to a large gravel parking lot with trails fanning out through the refuge. When Lisa drove us over to the other side, we ran into one of her students with her mother and grandmother. "You never know who you'll see out here," Lisa said, with a hint of pride.

That evening, we walked the few blocks from Lisa's place to call on Myrtle Jones. She is a former city council member and a founding member of the West Blocton Improvement Committee, the group that created the lily festival. Myrtle is in her nineties and lives with her daughter two streets back from Main. She had left the porch light on for us and met us at the door.

"I still remember the first time I saw the lilies," she said, lowering herself into a wide armchair alongside her frenetic black and white terrier, Maxie. "You know, some years the lilies come up in masses bigger than in others. The first time I walked down that dirt road and came around the bend in the river at Hargrove, it brought tears to my eyes," she said.

Myrtle moved to West Blocton in 1955 to teach social studies and Alabama history at the high school, and like Lisa, she enjoyed it well enough to stay.

She told us that the first festival was in 1990 at First Baptist Church. The next year it moved to the elementary school. Eventually the town bought the vacant department store for the lily gatherings and installed pews.

Myrtle mused on all the years and began totting off with pride the forms of recognition that the lilies have brought to West Blocton. "Let's see," she said. "*Southern Living* has given us a story, and several telephone books have featured the lily on the cover. There's a lily license plate you can get for your car from the State of Alabama, and Mercedes Benz even did a photo shoot with one of their cars at Hargrove Shoals. The ad ended up in *Playboy* magazine," she said, narrowing her eyes and smiling with mischief.

Larry Davenport told us later that the ad featured a dashing man in a tuxedo who was trout fishing in the Cahaba. "The lilies," Larry said, "though out of focus beyond the buffed-up SUV, were unmistakable. I only saw the page one time and then immediately destroyed the magazine." He laughed. "I teach at a Baptist school, you know."

Without getting up, Miss Myrtle directed us to a glass cabinet in her dining room that held commemorative plates and mugs that celebrated the lily. One plate, created several decades before the festival was established, featured a line drawing of the shoals on the front and an odd inscription in typewriter lettering on the back: *These beautiful and rare species of lilies are growing and blooming in profusion on Lily Shoals, in the Cahaba River in Bibb County, Ala. This particular variety of lily is found only in two other rivers in the world—the Amazon in Brazil and the Seine in France.*

"I don't know who made that up!" Miss Myrtle said, pulling off her glasses and wiping them. When we told Larry about the plate the next day, he said that was the kind of misinformation he had to deal with as he began his research for FWS.

Lisa reminded Myrtle that U-Haul had also chosen a bold image of a wildflower (though not the Cahaba lily) for the side of its rental trucks and trailers to represent Alabama and to draw attention all across the nation to the many rare species of wildflowers in Bibb County.

"Oh, yes," Miss Myrtle said, "and did you show them our town flag?"

"Yes," I said. We had seen it at the Lily Center. A wavy blue river ran the length of the flag with white block prints of the lily and its signature ribbon petals on one side and a row of red houses on the other shore. Donna and I had already bought lily T-shirts, commemorative enamel pins featuring a blossom, a tiny West Blocton flag, and a priceless community cookbook called *Coal Miners' Vittles*—all souvenirs from the Lily Center. In the cookbook, we found Lisa Buck's recipe for Cocktail Wieners, which (she notes in the directions) are a favorite of Dr. Larry Davenport. She also suggests that the recipe, which combines "a package of cocktail smokies, an 18-ounce bottle of barbecue sauce, and two tablespoons of grape jelly," should be heated on high for forty-five minutes and multiplied several times if served on Lily Day.

In the Lily Center we had pored over photos on the surrounding walls of prominent citizens and events in the life of the town, including yellowed newspaper clippings and formal color portraits of Miss Cahaba Lily through the years. We examined the student-created table displays on the biology of the lily. One exhibit advocated against the spraying of herbicides along the roads of Bibb County, to protect the rare native wildflowers.

As we prepared to leave Miss Myrtle, Lisa fished out the proceeds from our shopping spree at the Lily Center from the pocket of her hoodie and gave it to Miss Myrtle, who said she would deposit it first thing Monday in the West Blocton Improvement Committee bank account. We thanked her and bid her good night.

It was 1989 when Larry finished his research and filed his first report on the Cahaba lily with the FWS. At about the same time, the federal government declared a moratorium on adding any new species to the endangered list. Resources were apparently inadequate to manage any more plant candidates at the time. When the West Blocton Lily Festival got underway the next year, the publicity and interest that traveled across the state resulted in a rush of poachers, who began pulling up the lilies in the Cahaba and trying to sell them commercially.

"It became a real problem in the early nineties," Larry said. "I'd get these calls from people saying they had seen somebody with a pickup-truck-load of lily bulbs. 'What are you going to do about it?' they'd ask me." Larry stiffened his lips. "By this time, lily bulbs were being sold in the horticulture trade for $7.99 a bulb. I'd tried to grow them in a pot, and they don't do a thing," he said. "Unless you've got a living stream with sunlight, it doesn't work. I've tried to transplant them with very poor success. I tell folks to enjoy them where they are."

And for Larry, there was still more study to be done. He'd determined that the lilies across all three states only showed up along the fall line, where, as he put it, "the tough interior rocks meet the softer Coastal Plain rocks; that's where you get the shoals. So many towns in the Southeast were built at the fall line because that was the farthest upstream that boats could travel. They had to portage around."

But Larry still needed to determine what creature pollinated the Cahaba lily. "To be honest," he said, "I didn't know what I was doing."

A colleague at Samford set him up with an insect net and a killing jar. "I took a lawn chair and stuck it out in the shoals. I went out at dusk. I didn't have a radio, didn't want to listen to a ball game or whatever, I didn't want to disturb anything. At this point, all I had was folk information about what pollinated this thing. Some people swore it was bats. I didn't want it to be bats. I didn't want to deal with bats. Other people said butterflies, bees, hummingbirds. So I sat out there from six o'clock to eleven o'clock, five evenings in a row, and nothing was interested in this plant! The bats came out. They didn't care about this plant. Birds went home to roost. They didn't care about this plant."

Not long after those five miserable nights, Larry gave a talk to the local Sierra Club. "And this guy came up afterwards and said, 'I will find your pollinator.'" Larry took him up on the offer.

Dr. Randy Haddock was an entomologist who had studied at Cornell and was by that time earning his living as a medical researcher at the University of Alabama Birmingham Hospital. Getting out in the river was a great reprieve from his day job, and it didn't take him long to find his quarry. Randy identified the pollinator as the plebeian sphinx moth — a dramatic creature nearly as big as a hummingbird, with a wingspan as wide as the lilies.

Larry felt vindicated. He had hypothesized that the pollinator could be a moth, drawn after dark to the lily's brilliant blossoms and the profusion of sugar water they held.

"Everything pointed to a moth that has a tongue about two inches long," Larry said. "It hovers outside the flower and sticks its tongue in to get the nectar. That's what Randy found."

The two men soon arranged to meet at midnight at a Waffle House near the university. They drove down to the Cahaba near West Blocton. "And sure enough," Larry said, "at 2:30 in the morning, we're standing out there in the water, Randy is shining this light, and I saw these little pairs of red eyes going from one flower to another. It was surreal. And wonderful."

Randy Haddock wrote up the discovery for the CRS newsletter, "and it was huge news in Alabama," Larry said. "The next thing you know, Randy is hired as the research director for the Cahaba River Society, a job he's been doing for thirty years now."

Nearly a decade later, in 1996, Larry was asked to update his Cahaba lily

An elegant Cahaba lily opens on Hatchett Creek in an area known as Middle Lily Shoal in Coosa County, Alabama. Though each plant produces multiple blooms for nearly a month between Mother's Day and Father's Day, a single flower is at its peak for only a day. Used by permission of the photographer, Elmore DeMott.

report for FWS, and once again he traipsed across the South, discovering a few more populations here and there, including some small patches of lilies in tributaries off the main four rivers. He learned that the pipevine swallowtail—a handsome black butterfly with orange and purple spots—also pollinates the lilies, according to his sources in Georgia. Like many other botanists we interviewed for this book, Larry then brought up the unfortunate differences in the federal laws that protect animals versus plants.

"There is a human prejudice in favor of animals," he said. "People are much more interested in the big, brown-eyed Florida panther than, say, Furbish's lousewort, in Maine. The lousewort [*Pedicularis furbishiae*] is a rare perennial herb found only along the Saint John River in Canada and the United States." He continued, "It's a prejudice in both the law and perception. To protect something that doesn't move across the landscape is just not as interesting, I guess."

Larry's detective work topped out at seventy populations of lilies, including the two largest: at Landsford Canal in South Carolina, and at Hargrove Shoals on the Cahaba. The latter population, he estimates, is a half-mile long and a quarter-mile wide. "I'm afraid it would be impossible to take a congressman down there and say, 'Here's an endangered species,'" he told us.

Larry has surrendered any hope of a federal designation, at least for now. But with the establishment of the Cahaba River National Wildlife Refuge outside West Blocton, the biggest Alabama population is protected in a new way. The town of West Blocton also enacted an ordinance making it illegal to poach the lily. Lisa Buck told us that "people have been ticketed for messing with them."

Local enforcement makes the most sense, Larry said. "Recently I gave a talk in Helena, Alabama. They have a small population of the lilies down there on Buck Creek. This thing has caught on, and they are dedicating a park and calling it Cahaba Lily Park. The Helena town council wants to pass an ordinance making it illegal to *touch* the plants. Local ordinances have been very effective, and they come from our new knowledge about the habits of the plant."

Larry is quick to add, however, that when he comes to the end of his frequent talks about the lily, which he's been giving for thirty years now, he reminds his audiences about the preserves that have been established on both sides of the rivers in West Blocton and Landsford. "And then I ask them, 'What's upstream?' And the answer, of course, is Birmingham and Charlotte. That is where our problems lie. We have to do better upstream."

The CRS, headquartered in Birmingham, is a nonprofit membership organization that has been working since 1988 to protect and restore the Cahaba River Watershed and its rich diversity of life, which includes not only its globally significant freshwater biodiversity but also the diverse people of central Alabama.

The river is one of two main sources of drinking water for Birmingham. More than a million people, or about one-fifth of all Alabamians, largely depend on the Cahaba for their water.

Executive director Beth Stewart, research director Randy Haddock, and education director Gordon Black—all seasoned veterans of the work—joined us around a table in their offices. After years of fundraising, legislative and local government advocacy, and the marshaling of hundreds of volunteers for annual cleanup projects along the Cahaba, these three professionals have come to appreciate that the CRS needs to focus not just on the watershed but also on the "peopleshed" of residents and businesses along the Cahaba and well beyond to encompass the entire Birmingham metro area.

Many citizens whose lives intersect with and depend on the rivers confront significant barriers to environmental awareness and activism. For several generations now, young people have not been allowed or encouraged to play outside in nature. Some urban kids in Birmingham have never had their feet off the pavement for very long.

"The most important thing we do is simply get young people outside," Beth said. CRS sponsors experiences and environmental education activities year-round. La'Tanya Scott is the environmental science educator for CRS, a position she took in 2014 right out of college. She shepherds third and fourth graders on stream walks, takes older kids on canoe trips, and helps create and conduct service-learning projects for scouts, civic clubs, church groups, and schools. La'Tanya credits her parents with her lifetime love of the outdoors and water. The family often made camping, hiking, and fishing trips from their home in Temple, Georgia, when she was a child. Even though La'Tanya lost her father in a drowning accident when she was in grade school, she maintained her focus on the outdoors and earned a degree in environmental science at nearby Miles College. As an African American working with children in the Birmingham City Schools, La'Tanya is trying to give young people the same opportunities in the woods that she had as a child. "I had one student tell me he had never touched a tree," she told a reporter for the *Birmingham Times*.

Education director Gordon Black says students between fifth and ninth grade are still capable of being amazed by riverine ecology. He teaches students about the interdependence of fish and mussels, otters and raccoons, and turtles and birds.

"The teachers love the snails, and we have them participate in our water chemistry field experiments," Beth added. The program's teaching techniques are creative. Students divide into groups, and each has a different aspect of the river to follow on a map. The teams then teach each other about aspects of the Cahaba. CRS also engages artists to work with the river through music, graphic arts, poetry, and literature. These creative vehicles give voice and heart to the Cahaba River experience. Since 1996, more than thirty-eight thousand Alabama students have taken field trips to the Cahaba through CRS.

The Cahaba River Society's environmental educator, La'Tanya Scott, often conducts canoe trips and shoreline expeditions on the Cahaba with Birmingham elementary and middle school students. Used by permission of the photographer, Hunter Nichols.

Randy Haddock estimates that he has taken some eight thousand adults on paddle trips on the Cahaba since he joined the organization. In addition to the talks he gives about the benefits of lower-impact development and the establishment of tree buffers and bioswales along streams to reduce erosion and silt, he is working to raise awareness about wild taro—a large ornamental landscaping plant often called "elephant ears." Homeowners who have this plant in their yards inevitably thin out the plantings and tend to leave the corms, roots, and rhizomes at the street for pickup. "What they don't realize," Randy said, "is that the wild taro can wash into storm drains and easily take root along streams and rivers. The Coosa and Black Warrior Rivers have already suffered from the destruction of native plant species by wild taro invasion. Stopping wild taro from reaching the Cahaba lily populations downstream from Birmingham is now an urgent effort."

Among all these labor-intensive projects, Beth Stewart says that CRS, a predominantly white organization at its founding, is working double-time to represent the demographics of the region better in all of its programs, and on the staff and board. "Since our watershed was historically a destination for 'white flight' from the city, the Cahaba communities in the suburbs still tend to be associated with whiteness, even as they are becoming more diverse," Beth explained.

Today the population of the City of Birmingham is 62 percent African American, and the legacy of Jim Crow lingers in memory here. As Beth put it, "Nature is still seen among some folks as 'white space' or a place that is for whites, not for people of color. Some in the white population will actively question what a Black person is doing in nature, and Black people may feel unsafe when they meet white folks in the woods."

In the Deep South during enslavement, the woods and swamps were a conduit of escape for "runaways" and were often the site of brutal violence against people of color. The image of a grand oak with large, healthy branches might look like nature at its best to people like me, born white and privileged in the South of the 1950s. But for southerners of color, such a tree might conjure images of the multiple lynchings that took place throughout the region.

The National Memorial for Peace and Justice in Montgomery, which opened in April 2018, was designed to confront this legacy of terror in the collective history of southerners. As the founders of the memorial figure it, by 1950 more than 4,400 African American men, women, and children were hanged, burned alive, shot, drowned, and beaten to death by white mobs beginning in 1877, when Reconstruction began to fall apart.

A tendency to see danger in water was amplified by public policies excluding Black people from most beaches and public swimming pools in the first half of the twentieth century. "Parents who don't learn to swim can't teach their children to swim, so the aversion to water has become a generational and cultural phenomenon among many African Americans in the South," Beth said. When CRS invites Black students and teachers to experience the Cahaba River by getting in a canoe or kayak, some will say, "No. We just don't do that."

The Cahaba River is unfortunately associated with one of the major incidents of twentieth-century violence against people of color in Alabama. Highway 280 runs southeast from downtown Birmingham and crosses the Cahaba River at a historic bridge, where a group of white supremacists gathered beside the water to plot the bombing of the 16th Street Baptist Church on Youth Sunday in 1963. Four girls were killed in the blast. The perpetrators came to be known as "the Cahaba Boys," Beth explained. It was one of the most prominent tragedies of the civil rights era, and the wound has been slow to heal. Only one of the conspirators was sent to prison initially. It would be nearly forty years before two other men were convicted in the bombing.

Professor Anthony Overton, chair of the Department of Biological and Environmental Sciences at Samford, and a colleague of Larry Davenport, serves on the CRS board. Like La'Tanya, he grew up in a family that valued the outdoors and spent much of his youth hunting and fishing with his father outside Washington, DC, where he was raised. Overton credits his mother's influence, however, for his decision to study fisheries management and marine biology. She was a science teacher and charged Anthony with taking care of her classroom animals during summers when school was not in session. Anthony has lived in Birmingham for only three years and joined the CRS board at Beth's invitation, sparked by her passion for the work.

With his own children and as a volunteer for Jack and Jill of America, a leadership program for African American children and their parents, Anthony has always been a strong proponent of outdoor time for kids, and he is optimistic about the future. "In my opinion," he said, "our high school and middle school students are hands down more advanced than we were at that age. This whole Green movement has made them more aware of what they eat, for example, and their knowledge of sea level rise and what the word 'sustainability' means comes, in part, from the World Wide Web. They see it and learn more every single day."

Change is flowing in Alabama. It is ironic but somehow fitting that the term "lily white," once used metaphorically to signify the racial purity of white women in the South, now offers a new meaning, embodied in a rare flower that has become a galvanizing force to bring people together who might otherwise not have connected. "Everyone in the peopleshed depends on the Cahaba," Beth Stewart said, "and the survival of the Cahaba lily has become one important measure of our health as a city and state."

6

Morefield's Leather Flower

You can still find rural roads in the South where a particular flower will catch on among neighbors. You might drive by in the blooming season and see a dozen houses at a stretch where orange daylilies have been set out along the roadside. Or you might see heavy-headed peonies or dark green gardenias dotted with bridal-like blossoms banked against carports or garages. And so it is with clematis, an old-fashioned perennial climbing vine. Southerners will train them up a mailbox post or trellis or through porch rails to start a trend or to pay respect to their neighbors' plantings.

Some clematis varieties flower in spring. My grandmother favored the summer bloomers. These clematis—with broad, flat blooms—have four to seven sepals per blossom, depending on the variety. Colors can range from blue, deep purple, and pink to magenta. Some are variegated. The spiky stamens and/or pistils in the center of the flower are usually a contrasting color—often yellow. Some varieties have blooms as big as salad plates.

"Clematis" is from the ancient Greek for "climbing vine or twig." My family pronounced it "KLEM-uh-tis," and I've also heard some southern gardeners call it "kluh-MAH-tis." I was unaware until recently of a clematis variety that produces urn-shaped blossoms that hang down, as if the urn were being poured out. The flowers of these "Viorna" varieties remind me of the blooms on the wild ginger that we used to hunt by a creek in the woods when I was a child. My mother called these wild ginger flowers "little pigs." They are also known as "little brown jugs," and they hide under the much larger variegated leaves of the plant. To see them, you have to pull the green leaves back and gently brush aside the last season's leaf litter. Wild ginger, however, is a member of the birthwort family, and clematis—both the flat-petaled and jug-shaped varieties—belongs to the buttercup family, which also includes familiar flowers like columbine, hellebore, and monkshood.

The urn-shaped Morefield's leather flower (*Clematis morefieldii*) is a species discovered in 1982 at the base of Round Top Mountain in Huntsville, Alabama. It is a wild and endangered member of the buttercup family that is mostly found sprawling on boulders and shrubs where limestone outcroppings provide moisture and purchase for the vines that extend out as much as sixteen feet. The inch-long, reddish-purple flower buds are small and whimsical. Before they open, they look like miniature hot-air balloons. Eventually they crack open like a tulip, and the thick, neon-green, beveled edges of the leathery petals curl back sharply toward the hairy stem of the bloom. Their shape has been compared to a little bell. To see a prosperous vine in bloom ascending gray rock is like something out of a fairy tale. After flowering, pistils elongate into the feathery tails of the fruits—a puffy whorl of silky tan or golden-brown floss-like hairs. The discovery story of this otherworldly plant is a marvel.

In the late 1970s, when Jim Morefield was in high school, he was leaning toward the physical sciences as a career path. Meteorology, geology, and chemistry were all strong interests for him. Jim's grandfather, who lived in the Mojave Desert, was a physicist and a powerful influence on his grandson. Jim's father was an aerospace engineer, and that's why, in Jim's sophomore year of high school, the family suddenly picked up and left their home in what would later become better known as California's Silicon Valley.

Jim's father had been enlisted to work on the US Space Shuttle program in Huntsville, Alabama. Trading the high desert and California coast where he'd been raised for the humid, deep green South felt like no bargain. So when it was time to think about college, Jim found Deep Springs, a two-year school located on a working cattle ranch and alfalfa farm in a remote desert valley near Big Pine, California. He left Huntsville and returned to the West for college.

One of the smallest liberal arts schools in the United States, Deep Springs awards full scholarships to all students accepted and requires that they take full responsibility for themselves—what the college calls "self-governance"—as a means to cultivate independence, character, and decision-making skills. Students work on the farm year-round as they complete coursework for an associate's degree.

"I was definitely fascinated by the educational philosophy there—the integration of physical labor and self-governance," Jim explained. "I also liked the small size: a community of about forty, including students, faculty, and staff. It seemed like an opportunity to grow as part of something bigger than myself without being lost in a crowd."

"By happenstance, one of my last classes at Deep Springs was a course in ecology, taught by our natural sciences faculty member," Jim told me. "All I knew up to then about plants was that I loved the smell of sagebrush after rain." The four students in the ecology class, practicing self-governance, decided to tackle a field project to describe and map the vegetation communities found in Deep Springs Valley.

"To help make that work," Jim said, "I took it upon myself to learn as many plant species in the valley as I could, with no prior training along those lines. Fortunately, it was a relatively wet spring, and the desert wildflowers were coming out nicely. On several days I stuffed my day pack with the four thick volumes of *Abrams & Ferris's Pacific States Flora* (1923–60), walked out into the desert, sat down in the sand, and started identifying what was there. It helped that those volumes were well illustrated with accurate line drawings of the species."

Spending so many days with the field guide, Jim mastered botanical terminology, which made further identifications easier. "Certainly, this brute-force approach was not the most efficient," he said, "but it was effective, and helped me discover that I had some aptitude for botany, at which point I was irrevocably hooked."

Morefield finished his two years at Deep Springs and returned to Huntsville to live with his parents for a year and earn money for his next educational venture. In his free time, he began collecting and pressing plant specimens he found around Madison County. He gathered eight hundred specimens during that gap year, including samples of a leather flower from the eastern edge of Huntsville. "I had no clue at the time that it was anything new or unusual. It was just one of several specimens that I wasn't quite able to identify with confidence," Jim said.

Toward the end of the year, Morefield packed up all of his plant specimens and carried them to Nashville, Tennessee. He had been told that Dr. Robert Kral, a botany professor at Vanderbilt University and one of the most distinguished teachers and collectors of new species in the South at the time, could help him out. Like John Kunkel Small, Kral identified many new species and taught his students to do the same. A number of native species now carry Kral's name, in tribute to his work. At the time, Jim Morefield didn't quite appreciate Kral's stature in the field, but he had found his way to the right person.

"Dr. Kral was a total gentleman and scholar—gracious, welcoming, studious, patient, and eager to help a random young botany enthusiast he

A member of the buttercup family and sometimes called the Huntsville vase-vine, Morefield's leather flower (Clematis morefieldii) grows best in woodland shade. The deeply embossed, ovate leaves are spaced out along twisting vines, where delicate hairy blossoms emerge looking like little lanterns. Once open, they hang like jugs being poured out. Used by permission of the photographer, Alan Cressler.

had never met before," Jim told me. Kral looked through every specimen, confirming or correcting the identifications. "When he saw the leather flower," Jim said, "he immediately suspected that it was something new to science."

The budding botanist left a set of his specimens with Kral for further examination and headed to Northern Arizona University in Flagstaff, where he would earn a BS in botany and geology. In 1987, just as he was entering the PhD program in botany at Rancho Santa Ana Botanic Garden and the Claremont Graduate School in California, Morefield got word that Dr. Kral had published the name, description, and his own superb drawings of a new species of clematis from northern Alabama in the prestigious *Annals of the Missouri Botanical Garden*. Kral had named it *Clematis morefieldii*, writing: "Mr. J. D. Morefield, a careful and perceptive student of botany, is gratefully acknowledged. During the few years of his adult residence in Huntsville he developed an excellent personal herbarium which adds much to our information about the flora of northern Alabama.... The Clematis is therefore named in his honor and as a reminder that he is missed back east."

What a grand entrance to graduate school!

"It was one of a few pivotal experiences that motivated me to continue in that field of study to this day, for which I am ever grateful," Jim told me. He is now supervisory botanist for the Nevada Natural Heritage Program in Carson City. Dr. Kral has long since retired from Vanderbilt, and in 2012 he became the fifth person to be inducted into the Tennessee Native Plant Society Hall of Fame.

Even though I knew that by mid-June it was probably too late to see Morefield's leather flower blooming in Alabama, I set out very early one morning for Huntsville. From the middle of North Carolina, I drove west on Interstate 40 for four hours. Then I opted for a less traveled four-lane— US 74—to carry me into the Smoky Mountains and beyond, to Alabama. My goal was to arrive by late afternoon at the Keel Mountain Preserve, now managed by The Nature Conservancy of Alabama, acquired specifically to protect the endangered Morefield's leather flower. The endangered Cumberland rosinweed (*Silphium brachiatum*) and the rare limerock arrowwood (*Viburnum bracteatum*) also grow on the preserve. A dramatic sinkhole where a waterfall flows over a limestone cliff is one of the main attractions of the preserve, and the creeks flow vigorously on the property. Keel Mountain is the largest known site for Morefield's leather flower, with an estimated four hundred vines within a quarter acre.

Eight hours into my drive west into unknown territory, I was relieved to find that the destination seemed not so terribly remote. I appeared to be coming into an outer ring of Huntsville suburbs, where small farms and clusters of midcentury ranch houses with acreage stretched out in a congenial mix. Above, afternoon thunderheads were forming. If you've ever flown over this section of Tennessee and Alabama, you've probably seen how the Blue Ridge and Smoky Mountains on the east side are an impressive collection of randomly arranged, rugged peaks. But beyond the Smokies, the topography becomes a series of parallel, flat-topped ridges, straight as furrows, running more or less northeast to southwest. Huntsville is flanked by this view of straight-line plateaus. Admiring it, I wondered more than once, Would my map app remain connected to a global positioning system server this far out of the city and hold the signal to my destination?

Then the bottom fell out of the sky. Rain drilled the car roof and hood. It fell in big drops so dense I couldn't make out street signs until I was right up on them. Cars passing in the other direction threw up sheets of water that my wipers could hardly rake aside. I turned on my headlights and slowed down. I was thankful that the arrow on my phone screen kept moving on down the road to the next turn and the next. Eight hours of driving and now this washout. I was unnerved. Several turns later, I was nearing the destination on the map and the rain abruptly stopped. The sun popped out. I was still anxious. I did not see any sign of the preserve. I passed the street number listed online. I stopped and turned around. Heading back from the opposite direction, I could see a narrow break in the forest that led to a clearing—a gravel parking lot in a sparsely developed neighborhood. By now the ground was steaming. Little rags of fog lifted from the pavement. No other cars were parked on-site, so probably no other hikers were in the preserve. A bulletin board mounted under a sturdy shelter offered a trail map and warnings not to veer off the path.

Going into a forest alone is not my usual practice. I had actually bought an air horn before I left home, anticipating this part of the trip. But it wasn't likely, I realized, that anyone would hear the horn if I used it to scare a bear or to summon help. The description of the preserve said it is surrounded by private land and to stick to the trail. You bet I would stick to the trail. The trees shed water on my notepad and my scribbling blurred. The rocks in the middle of the trail were slick. The creek alongside me was running fast. I moved along, noting the poison ivy and looking for viny plants. Except for the dripping from the overstory, the place was so quiet that nothing could sneak up on me, I told myself. Exposed roots threaded through the limestone outcroppings as I climbed slowly.

A few plants bloomed in a thick, hardwood forest of oak, smoke tree, hickory, and cedar. My contact at the Huntsville Botanical Gardens had already told me that the blooming period was over for her Morefield specimens in cultivation, but I might see the swirl of silken threads left where the seeds would be forming. The seeds of this clematis can remain dormant in the ground for a year or more, she had said.

I took photos of what I later identified as bottlebrush grass (*Elymus hystrix*), with bristled white spires rising festively like fireworks from the ground. I also saw buttonbush (*Cephalanthus occidentalis*), a member of the coffee family with perfectly round balls and spikes that suggested a miniature pincushion. There was leafcup (*Polymnia canadenisis*), with its white blooms; yellow St. John's wort; and a most elegant little fern that seemed to grow right out of the limestone. The *Trillum cuneatum* I saw—also known by the diverse common names of little sweet Betsy, purple toadshade, or bloody butcher—was amazing, with its three, camo-colored green leaves and a single magenta bloom rising from the center.

I stepped carefully, mindful of snakes that love rocks, and continued to look for vines. I spotted two that might have been clematis to my un-trained eye, but no swirls of silken seedpods. I photographed the vines: one with a narrow, arrowhead leaf; the other with a wider leaf that was deeply veined, almost embossed, it seemed. Once again I felt the profound enormity of the natural world and began to think about Small and Heller, wandering in the wild South when wilderness was much deeper than today. The phrase "needle in a haystack" came to mind. I despaired, and at the same time began to appreciate the task of the taxonomist: the sheer volume of inspecting, guessing, comparing, detailing, drawing, identify-ing, or—in the case of the hitherto unknown—naming.

In the end, I walked thirty minutes in, and a little more quickly out. I had not found Morefield's leather flower or the sinkhole. I had not twisted my ankle or been bitten by anything more dire than a mosquito or two, and I had not fallen on my backside. I was glad to get back to the car, feeling a bit overwhelmed.

I had dinner with friends and spent the night in Huntsville. The next day arrived with blue sky and sunshine, and I was off to the Huntsville Bo-tanical Garden, an impressive facility on 112 acres near the US Space and Rocket Center. Both the Garden and the Rocket Center were developed as destinations for visitors on land adjacent to the Redstone Arsenal: the US Army's historic missile development and testing facility and one of Ala-bama's major economic engines. Huntsville Botanical Garden's primary attraction is a seasonal, open-air butterfly house—the largest one in the nation. I was to meet the curator of horticulture, Tracy Cook, in the taste-ful, white-columned Propst Guest Center. Opened in 2017, the center is the focal point of the drive into the grounds. The expansive building has a wide veranda with rocking chairs, a glass-domed conservatory, a restau-rant, and a grand hall for public events.

Tracy and I found each other quickly and wove through the schoolchil-dren who had just arrived on buses and were clotting up the lobby. We made a fast getaway out the back to Tracy's golf cart and headed for the Matthews Nature Trail. Some seventeen specimens of Morefield's leather flower (also called the Huntsville vase-vine) were planted here more than thirty years ago. Tracy told me that the provenance of their plants is "sketchy." They were a gift to a garden volunteer from a friend who lived in Tennessee. Thirty years ago, however, no one had identified the *Clematis morefieldii* in any state besides Alabama, though sites have since been identified in Tennessee and Georgia.

As we sailed over a pristine pebbled path through the formal gardens, Tracy gave me a bit of her history. Her family settled in Huntsville when she was six and then moved when she was twelve to rural Arab, Alabama, where her mother kept horses. They left town the same year the development of the Huntsville Botanical Garden was announced. Tracy was bummed. Even then she had a hunch that she wanted to be a part of the garden's development. She had grown fond of the green spaces and mature gardens around Huntsville. Living in rural Alabama just wouldn't be the same, she felt.

As a teenager, Tracy began cultivating mail-order roses and "tinkered with planting a garden." Because of limited family funds to support her going to college, she joined the Marine Corps after high school and did basic training at Parris Island, South Carolina. Later she trained and worked as an avionics electrician at New River Marine Air Station in eastern North Carolina. "I missed Huntsville during those years," she said, but she kept growing plants in pots. "I probably overwatered them," she said, and smiled. Now in her forties, Tracy still has the disciplined demeanor of a marine. She was wearing a no-nonsense gardening outfit made of a sturdy fabric designed for outdoor work.

When her commitment to the Marines was finished, Tracy enrolled in the horticulture program at Auburn University on the GI bill and graduated in two years. She then worked in landscape design in north Georgia for five years. When a job opened up at the Huntsville Botanical Garden in 2011, she was so eager to apply that she delivered her résumé even though eleven tornadoes were tearing across northern Alabama that day. She got to Huntsville and got the job. Working as assistant curator for horticulture, Tracy earned a master's degree in plant and soil science at Alabama A&M, finishing in 2018. Her master's thesis focused on Morefield's leather flower.

"The species caught my attention in 2014, when a local genome lab confirmed that the DNA sequence from one of our plants here matched that of *C. morefieldii*," she explained as we rolled into the native plant section of the garden. The DNA lab — Hudson Alpha Biotech Institute in Huntsville —

happens to be one of the biggest genomics researchers in the world. "Hudson Alpha," Tracy said, "runs a summer academy for local high school juniors and seniors who collect samples from our native Alabama plants here in the garden. The students learn how to extract the DNA, and then the lab does gene sequencing to prove the species. It's a great experience for the kids."

When the clematis was sequenced successfully, it made the newspapers, and the students in the summer academy had their names permanently recorded in a database as the researchers who completed the gene sequencing of the species, which is now identified by a unique bar code. Scientists around the world can view these DNA sequences and compare them to other, similar species through a worldwide reference library known as the International Barcode of Life. It's a mechanism for identifying species and their characteristics that's completely different from those archival sheets of brittle, pressed plants and flowers from the nineteenth century that we saw at the New York Botanical Garden.

Compelled by the rarity and fragility of Morefield's leather flower, Tracy devised a master's thesis project. She put to use the GIS (Global Information System) mapping and design skills she'd acquired in graduate school to create "a spatially explicit predictive map of possible sites" where the Morefield leather flower might occur in the wild, in addition to the known sites in the immediate region.

To create the map, she plugged in all the known geographic coordinates for the plant, using data from the Alabama and Tennessee Natural Heritage Programs. She then analyzed the topographic parameters of those sites to draw some conclusions about the kind of habitat where the leather flower readily thrives. As she explained to me: "The species distribution model identified a highly suitable habitat on a particular mountain near Huntsville about three-quarters of a mile up a specific trail. So we went there, and we counted over forty stems of the vine in that location!"

"Very cool," I said.

There were likely more sites in the vicinity, Tracy added, but the terrain was too rough to search without climbing gear. Still, the site was not a known or recorded location for Morefield's leather flower at the time of her research.

Tracy stopped the golf cart near a semishaded bed where several clematis vines were nestled among other Alabama natives. I saw the whorls of hairy threads where the leather flowers had been and noted that the leaves and stems were similar to the vines I'd seen the day before on my lone hunt in the dripping woods of the preserve. I took pictures to com-

Tracy Cook, curator of horticulture at the Huntsville Botanical Garden, picks up a strand of Morefield's leather flower vine (Clematis morefieldii) growing after the plant's blooming season along the garden's Matthews Nature Trail. Photo by Georgann Eubanks.

pare the leaves with the images from the day before. What I'd seen at the preserve definitely had a different shape, I realized. For a novice like me, however, it would be easy to confuse the vines.

Tracy and I made our way over to a second group of the Morefield's leather flowers, in a sunnier location beside a running creek. These plants were clearly happier than the ones in shade. I hated that I'd missed the blooming time.

We got back on the cart to ride to the nonpublic section of the garden. Tracy took me into a greenhouse where seedlings of the leather flower were growing. Huntsville Botanical Garden is partnering with ABG on native plant cultivation and conservation, and Morefield's leather flower is a focal point here.

I asked Tracy about the lack of support and appreciation these days for horticulture as a field of study, noting how a number of once-dynamic horticulture programs at colleges and universities in the South and elsewhere had been folded into other academic departments or eliminated owing to underenrollment, decreasing state resources, and faculty retirements.

She shook her head. "Horticulture is still seen as stoop labor," she said. "It's not always perceived as science, but it is science. It's applied science. It's boots on the ground. Researchers who spend their days in a lab with tissue fragments of a plant often have to depend on us to know enough to recognize when something is going wrong with the whole plant."

For her part, Tracy has continued following the literature about the ongoing threats to *Clematis morefieldii* in the wild. Besides being hurt by industrial and residential development, she told me, the plant is plagued by destructive deer, mealybugs, and mice.

"Rodents find the seeds very tasty," Tracy said, "and the plant will abort flowers if the shade gets too deep where they are growing." She noted that there are actually four known kinds of leather flowers in Alabama, but they don't tend to cross-pollinate, because their bloom periods don't overlap. What delineates one from another, she said, is how pollinators interact with them: four different types of bees are involved. "But I'm not a bug gal," Tracy said, grinning. "Pollinators definitely affect leather-flower livelihood, and pollinators get thrown off by climate change, including the extremes of drought and rain that we've experienced lately," she said.

So many challenges from so many factors. We were still talking about preserving biodiversity in the South when Tracy got me back to my car in the parking lot. "We have to conserve habitats, not just the species," she said finally. "We don't know where the next cure for cancer is coming from. The world's wealth is really in our plants."

I thanked her for her time, and she promised to send me her master's thesis and some pictures of the Morefield's leather flower from when they were blooming in the garden.

My next stop was a little more than an hour northeast of Huntsville. I'd arranged for a visit to the Sewanee Herbarium, at the University of the South in southeastern Tennessee. *Clematis morefieldii* was discovered there by Todd Crabtree, state botanist for the Tennessee Natural Heritage Program, who monitors the plant's locations in this corner of the state. When I inquired about his findings, Crabtree wrote to explain that in the most recent data compiled for the five-year update required for every federally endangered species, there were now twenty-two separate known populations of Morefield's leather flower across Tennessee and Alabama, but "only four are either excellent or good estimated viability," and "eight have a poor estimated viability." Having said that, Crabtree offered this analysis: "Many of these populations are on public lands, and we have started managing the habitat at some sites to benefit the species. Based on my experience over the past ten years, I think the prospects are good for the species."

I had never been to Sewanee, where the University of the South sits at the south end of the Cumberland Plateau, high atop Monteagle Mountain on thirteen thousand acres known by the academic community as "The Domain." The campus feels, in many ways, like a private island in the sky, overlooking the central Tennessee Valley to the west. The university opened in 1868 and has a distinguished history as a small liberal arts school. It is the sole beneficiary of playwright Tennessee Williams's estate and has furthered the careers of many southern writers. In addition to its bright literary burnish, the university has recently set the goal of becoming a national leader in environmental studies and sustainability.

The main buildings of campus, most of them reddish-gray sandstone, are surrounded by faculty cottages, larger Victorian houses, and handsome log cabins. A small commercial village off the main highway set a tone like that of my college days in the 1970s — the earthy-lentil-soup-and-sprouts-on-your-salad era. After a quick lunch at a retro hippie café, I hunted down the Sewanee Herbarium and met the director, Professor Jon Evans, whose specialties are botany, ecology, and conservation biology. He was working in the lab that day with two herbarium undergraduate fellows — rising junior Lillian Fulgham, who grew up in the woods of Mississippi and was homeschooled, and rising sophomore Angus Pritchard, from Decatur, Georgia, a dedicated birder who already had plans to earn a PhD in ecology.

It was quickly apparent that both Lillian and Angus are keen advocates of the preservation of southern biodiversity, especially on the Domain. Through her fellowship, Lillian was assisting with research on the genetic structure of the rare Appalachian hill cane that still grows on campus property. She also played violin in the school symphony and trombone in the jazz ensemble.

Angus was studying the impact of climate change on a forest that floods annually on the Arnold Air Force Base, twenty-five miles northwest of Sewanee. Angus told me that he had recently become an avid user of a phone application called Seek. Created by the California-based nonprofit iNaturalist, the app uses a phone's camera to identify plants in the wild with something like the facial recognition software developed by Google, which is linked to iNaturalist's deep and growing database of worldwide flora and fauna. (It's an amazing technology that I tried out the day before to help me identify some of the blooms in the Keel Mountain Preserve, though it didn't work with all the plants I aimed to identify because, as Jim Morefield later explained to me, it only works well for species that are more common and frequently observed—at least in its present stage of development.)

The Sewanee Herbarium Fellows Program underwrites a small group of students who work on campus in the summer doing biological research. These projects can even be continued after graduation for a year of post-baccalaureate work leading to publication. The fellows also organize and conduct campus tours to increase plant awareness among students and visitors at Sewanee—a direct effort to counteract plant blindness.

Over the years, Jon led the effort to identify more than 1,130 distinct plant species on the Domain—now in an archive called Vascular Flora of the University of the South, which has been digitized in a database and published in *Castanea, the Journal of the Southern Appalachian Botanical Society*.

Herbarium curators Mary Priestley and Yolande Gottfried were also instrumental in establishing the collection, which Sewanee claims gives it the highest documented plant-species diversity of any college campus in the United States.

Jon Evans was raised in Maryland and studied at Cornell as an undergraduate. He earned his PhD in botany at Duke, where he did research on the barrier islands of North Carolina. He has been at Sewanee since 1994 and was responsible for a campus-wide initiative to weave sustainability and environmental stewardship throughout the curriculum and through a defined set of university conservation practices. He was the architect of the first Sustainability Master Plan for the university and created a Landscape Analysis Lab where students and faculty can use GIS mapping technology for their research and teaching. Jon also represents the university in regional efforts to protect the Cumberland Plateau — a 300-million-year-old eruption of ridges, cliffs, gorges, and bluffs layered with coal, limestone, and sandstone. The entire plateau is home to extraordinary biodiversity and features long stretches of contiguous forest as it runs from eastern Kentucky through Tennessee and into northern Alabama and northwest Georgia. Near Sewanee, the rocky composition of the plateau where Interstate 24 cuts through it looks like an old-fashioned Appalachian stack cake.

As I chatted with Jon and his students in the Herbarium office, curator Mary Priestley brought in a pressed specimen of *Clematis morefieldii* from the Sewanee collection. It had the signature whorl of silky threads where seeds develop after the blossom has fallen away. The notes on the mounting sheet revealed that the sample had been confirmed by Dwayne Estes as *Clematis morefieldii* in June 2005. The citation read: "Woods south of TVA powerline right-of-way south of the War Memorial Cross, 1400' elevation." (The War Memorial Cross is a Sewanee landmark on the bluff overlooking a sharp drop to the valley below. It commemorates World War I casualties from the region.)

In 2003, when they were graduate students at the University of Tennessee, Dwayne Estes and Chris Fleming discovered the first site of Morefield's leather flower in the state. It was in Franklin County, Tennessee. Later Estes helped to confirm other plant specimens with Jon Evans in the creation of the Domain flora collection. He is now a prominent botanist at Tennessee's Austin Peay University in Clarksville and cofounder of the Southeastern Grasslands Initiative, which is discussed in chapter 10.

Jon explained that Tennessee's state botanist, Todd Crabtree, had continued to refine the geographic parameters and conditions that predict where Morefield's leather flower might be located in the region. Like Tracy Cook in Huntsville, Crabtree had built a model using GIS mapping to predict precisely where he would find the species in Grundy County, and the plant was indeed there. Dwayne Estes and the late Tom Patrick (then state botanist of Georgia) and two other plant hunters had also found *Clematis morefieldii* on Pigeon Mountain in Walker County, Georgia, in 2015, adding a third state to the provenance of the endangered species.

Relying on his experience with leather flowers and pitcher plants in

Lilian Fulgham and Angus Pritchard, undergraduate fellows at the University of the South, in Sewanee, Tennessee, along with Jon Evans, a biology professor there, hold an herbarium specimen of Clematis morefieldii, *the only endangered species found on campus. This pressed sample of the Morefield's leather flower exhibits the whorl of silky threads that develop after the blossoms fall away. Photo by Georgann Eubanks.*

the region, Dwayne Estes has made an argument that botanists cannot rely solely on dried herbarium specimens to tally the true magnitude of biodiversity in the South. Using *Clematis morefieldii* as an example, Estes noted that the vine grows much longer than herbarium paper, which is typically 18 inches by 11.5 inches. The true shape of the plant's three-dimensional bloom is also flattened by the plant press, and the distinctive colors and scents of the blossoms are lost in the drying process. Such sensory factors have not always been documented by collectors, either. As Estes put it: "Once their parts become smashed and dried, the various species often look alike, and differences that seem to exist when viewing living plants side-by-side in the field or garden seem to break down in the herbarium. This has led many botanists to do a lot of 'lumping' in some of these groups."

By lumping, Estes meant the conflation of potentially distinct species into one taxonomic name. He argued that in the 1800s, northern botanists such as Asa Gray and others, who occasionally traveled to this region, depended heavily on herbarium specimens that had been sent to them by southern botanists. These local collectors sometimes lacked the stature of their northern counterparts. The result, Estes wrote, is that "Northern botanists would look at specimens [from this part of the country] that resembled species common in the Northern states, and they would lump them together as one species." In his work, particularly with leather flowers, Estes found discrete differences among specimens that might qualify them as unique—possibly up to six additional species.

"One notable exception to the Northern lumping botanists," Estes wrote, "was John Kunkel Small of Columbia University." Small was often criticized for declaring specimens he found in the region to be completely new species, as documented in the 1933 edition of his manual, *Flora of the Southeastern U.S.* Estes wrote: "Many of Small's species that he recognized in 1933, which were subsequently disregarded by many later botanists, are now being upheld by science as good species and are being resurrected. Could something that has been generally thought to be one species be divided into two or more species?"

Botanists now have much better tools to distinguish among specimens—historic and living. The botanist's capacity to sequence plant DNA has changed the game, even if very old specimens may not read correctly because of physical degradation. According to Estes, the historic underestimation of biodiversity in the South could have other consequences: "We may be failing to protect critically endangered species that we don't realize exist."

Clematis morefieldii is the only federally endangered species discovered so far on the Domain, though Jon and his students have been adding new species to Sewanee's Flora database as they continue to find them. The *morefieldii* plants on the Domain are not easy or safe to visit, they told me. The plants grow best on steep rock outcroppings far down the mountainside from the university's classrooms.

The plant's preference for south-facing rock ledges along the Cumberland Plateau has more than once put the species in the path of new rock-quarrying operations, particularly around Huntsville. Management decisions for the conservation of the species will require that botanists continue research into the plant's biology and reproductive processes to protect it in these situations.

As Jon Evans was quick to add, the mechanism of seed dispersal in *Clematis morefieldii* is still only partially understood. Sewanee's work to engage students in these kinds of thorny questions through a popular course in conservation biology is a positive model. Jon said he believes that technology, which is often seen as a hindrance to students' appreciation of the natural world, can actually enhance an experience in the wild.

Tools such as the Seek phone app that Angus Pritchard was using the day that I visited encourage students to identify plants, insects, and animals in the wild and to maintain a running record of their findings on their handheld devices. "Many students who are interested in biology tend to know their birds, but they don't know plants," Jon told me. "But they will discover them here," he said with sly confidence. Perhaps that's why, a month after our visit, Professor Evans was named Conservation Educator of the Year by the Tennessee Wildlife Federation.

7

Michaux's Sumac

 Desperate times call for desperate measures. Mincy Moffett knows. As a rare-species botanist with the Wildlife Resources Division of the Georgia Department of Natural Resources (GDNR) for many years, he has been tracking across the state some 30 species listed as endangered by the federal government and another 155 species protected by state law. According to the Flora-Manager database kept at the North Carolina Botanical Garden Herbarium, Georgia ranks sixth in vascular plant diversity in the United States. Keeping up with rare plant species is a big job, but thanks to his background in business, Mincy has a knack for making science accessible to the public and for cleverly promoting the value of endangered plants.

A decade ago, when the little-appreciated Michaux's sumac (*Rhus michauxii*) had dwindled to only two small and struggling populations in Georgia, Mincy cooked up a Valentine's Day project to bring attention to the plight of the plant. With colleagues from the Atlanta Botanical Garden (ABG) and the State Botanical Garden of Georgia, he staged a romantic encounter in the wild between specimens of Michaux's sumac, drawn from male and female populations that had been living separately in Georgia, perhaps for as long as a century. *Rhus michauxii* is a dioecious shrub, meaning that almost all flowers on individual plants have either male or female reproductive parts, but not both (although a 1991 study found that some stems on an individual plant may be capable of changing sex). Regardless, flowering plants of both sexes must be in proximity to reproduce by insect pollination. Apparently, only clonal versions of the plants were growing in Georgia, albeit poorly, by extending rhizomes (below-ground stems capable of making new roots and shoots) at both locations. Without an opposite-sex plant nearby, the species' genetic diversity also declines, which is what had put the plant on the precipice of extinction among the state's surviving populations in the wild. As Mincy explained, the last known stand of dwarf sumac females survived under a water tower in Newton County, southeast of Atlanta. Eighty miles away, only five male specimens of the species were left on a forested bluff overlooking the Broad River in Elbert County, near Georgia's northeastern border with South Carolina. After a controlled burn on the bluff, Mincy's team was able to boost the population of males there to twenty-five specimens, so the time was ripe for matchmaking.

With a newspaper reporter and photographer in tow, Mincy and Jenny Cruse-Sanders, who would eventually become director of the State Botanical Garden of Georgia, led a team carrying female stems that had been cultivated by ABG from cuttings taken at the water tower. They planted the females alongside the male plants on the river bluff. "Let's hope this turns into a torrid romance," Mincy told a reporter from the *Athens Banner Herald*.

The dwarf sumac was discovered by André Michaux in 1794 in Mecklenburg County, which is now the most urban county in North Carolina. (Charlotte is its county seat.) Charles Sprague Sargent, the first director of Harvard's Arnold Arboretum, named the sumac for the omnipresent French botanist in 1803. A perennial shrub in the cashew family, it is also sometimes called false poison sumac, distinguishing it from poison sumac (*Toxicodendron vernix*), which can cause a contact rash like that caused by poison ivy and poison oak. The South's better-known native is the winged sumac (*Rhus copallinum*), especially beloved in the Appalachian Mountains for its fiery red autumn foliage and berries. *R. copallinum* can reach treelike heights of up to thirty feet, though it commonly stands at ten to twelve feet.

Michaux's sumac is distinguished by its extremely hairy stems and green serrated leaflets and by the elegant fountain of tiny greenish-yellow flowers that emerge on the female at the center of the shrub in June, surrounded by symmetrical whorls of branches—a vision that botanists call "densely pubescent." The blooms on the female plant eventually transform into a cone-shaped profusion of red berries in the fall. The leaflets soon follow by turning crimson. As a sun-loving plant that reaches a maximum height of only about three feet, the dwarf sumac requires an open habitat, which has contributed to its demise in the dense forests of the South. The federal government declared it endangered in 1989.

By the time a recovery plan was developed for the species in 1993, twenty populations, including the last one in Florida, five in Georgia, three in South Carolina, and eleven in North Carolina, had been lost. The lack of genetic variation among the surviving populations seemed to be the result of the reduced range of the plant in the region and was therefore a continuing threat to species survival. Other factors in the dwarf sumac's decline included loss or degradation of habitat from commercial, residential, and industrial development, along with encroaching agriculture and silviculture, which can bring off-target herbicide drift.

Half of the remaining twenty-one populations discussed in the recovery plan were located "on roadsides or on the edges of artificially maintained clearings," where they still received adequate sunlight. Scientists believed that the suppression of natural fire in landscapes where the sumac once thrived was the initial circumstance that launched the decline.

This hypothesis was confirmed when botanists discovered the largest remaining population of Michaux's sumac on Fort Pickett, a military base in the Virginia Piedmont. There, soldiers conducting training maneuvers using live munitions and artillery regularly ignited small fires that eliminated underbrush. Bombing exercises using even larger ordnance opened up the tree canopy and eliminated all but the fire-tolerant species, allowing the sumac to prosper.

In the Sandhills of North Carolina, Fort Bragg, the world's largest military installation, is also home to several endangered species. Unlike at Fort Pickett, these plants are now protected from any activity that would damage them, including military exercises. Instead, parts of the base that harbor fragile plants are managed with controlled burns during the growing season between April and October in a three-year rotation. Fort Bragg's rarest plants—in addition to Michaux's sumac—are American chaffseed (the subject of chapter 10), Venus flytraps (with locations that are a closely guarded military secret), and rough-leaved loosestrife (*Lysimachia asperulaefolia*), a member of the primrose family that bears dainty yellow flowers that smell like bubble gum. Twenty-eight of the fifty-six populations of this loosestrife known in the world live at Fort Bragg. The base's Endangered Species Branch is staffed by dedicated scientists who also monitor and protect the rare St. Francis satyr butterfly and the red-cockaded woodpecker.

Once the Valentine's Day plantings of Michaux's sumac proved successful, Mincy and his colleagues continued a regular regimen of controlled burning to open up the overstory and keep the underbrush cleared on the bluff overlooking the Broad River. Mincy offered to take us to this site, which the Army Corps of Engineers leases to GDNR to manage for conservation and recreation. We hoped we might witness one of their prescribed burns, but the weather did not cooperate. Three days of torrential rain had rendered the woods dripping wet during our February visit to Georgia. Still, Mincy agreed to squire us to the Lower Broad River Wildlife Management Area in Elbert County to see the sumac up close.

On the Auburn University website I'd already read an article about Mincy that amazed me. In his late twenties, before earning his PhD in plant biology, he had hit a low spot in his life. He'd majored in economics at the University of the South, earned an MBA at Georgia State University, and then landed a plum job with Georgia Federal Savings and Loan. But he was miserable. "I had a career that I hated and two degrees that I didn't want to use anymore, so I thought, 'Man, I've got to make a change,'" he told the Auburn alumni magazine.

Mincy left Atlanta in 1987 and went to live with his brother in Washington, DC. There he realized his heart was in the natural world. Using his business credentials, he soon acquired a desk job with Greenpeace USA, a nonprofit environmental watchdog. He eventually became the organization's administrative director while also volunteering for several nonviolent protest actions.

In 1989, in the midst of new global concerns about the destructive impact of chlorofluorocarbons (CFCs) on Earth's ozone layer, Mincy helped hatch a plan to hang a huge banner on a DuPont-owned water tower along Interstate 95. The banner highlighted the corporation's status as the world's number-one producer of CFCs, under the brand name Freon. Having managed a few rock-climbing expeditions in high school and college, Mincy was comfortable enough with heights and technical ropes to scale the 180-foot tower, hang the 65-foot-tall banner, and spend three days camping on the top in protest. Television reporters hovered nearby in news helicopters filming Mincy and his fellow Greenpeace activists. They gave media interviews using bulky, first-generation mobile phones that Mincy said "looked like a hybrid between a World War II field radio and a modern walkie-talkie."

On his last high-profile protest, in 1994, Mincy helped lead a team who disguised themselves as elevator repairmen and ascended to the roof of the *USA Today* building on Pennsylvania Avenue in Washington, DC. There, Mincy and his daredevil mates bravely hopped across the gap between buildings to land on top of the World Bank. They used the window-washing arms permanently mounted on the roof to rappel down the building, while employees were in the middle of celebrating the fiftieth anniversary of the bank's establishment. Staff watched openmouthed from their office windows as Mincy's team unfurled a banner protesting the organization's complicity in the funding of deforestation projects, river impoundments, and the forced migration of Native populations from their homelands. The banner featured a cartoon monster called "World Bankenstein" wielding a chain saw. As Mincy was quick to point out, this gambit took place only a few blocks from the White House and seven years before the 9/11 attacks on the World Trade Center, which heightened security measures and lowered tolerance among law enforcement for such actions.

"Later on," he said, "we would have been shot."

By this time, Mincy had met a woman at Greenpeace who would become his wife. Now in his thirties, he longed to go back to school in environmental science, which would require a lot of catch-up courses in biology, chemistry, and physics. He and his wife returned to Georgia, where he prepared to apply to graduate school. He ended up at Auburn to study for his PhD. His thesis committee included the esteemed Robert Kral, by then an emeritus professor at Vanderbilt and the man who had helped Jim Morefield during his leather flower discovery. Mincy had finally found his niche.

At our appointed hour, Mincy and botany technician Morgan Bettcher arrived in their big gray four-wheel-drive state trucks for the forty-mile drive from Athens (where we were staying) to Elbert County, to see the sumac. Mincy wore a starched gray state-issued GDNR shirt over a long-sleeved turquoise T-shirt and heavy-duty work pants. It was a chilly morning. His hair, now going gray, still had plenty of sun-bleached strands on top that set off his intensely blue eyes. Morgan, originally from Auburn, Alabama, had come to Athens from the University of North Carolina, where he had completed his undergraduate degree on a full scholarship. He began his graduate work in Odum Ecology School at the University of Georgia but left after one semester to begin a career with GDNR doing water-quality testing with the Environmental Protection Division. Now, at age twenty-eight, he is the first full-time botany technician on Mincy's team in the Wildlife Resources Division. "He's definitely taken to the plants," his boss said. "He's been a huge asset." Morgan, a curly-haired man with chiseled features, wore the dark-blue version of the state-issued shirt and a fleece jacket. Both men had laced on their well-worn, heavy-duty field boots.

Donna took a seat in Morgan's truck, and I hopped aboard Mincy's. I immediately felt like I was with an old friend. Mincy's passion for his work was palpable: the space behind the bench seat in his vehicle was stuffed with every conceivable item—tools, tarps, fire gear—that he might need in the field.

Though Mincy and I are a few years apart in age, we soon figured out that we had grown up in the same Atlanta world of privilege, with a sense of endless possibility in the 1970s. He went to a private Catholic military academy for boys that was both a football and basketball rival of my public high school. We moved in adjacent circles. Mincy's father was a lobbyist for the Medical Association of Georgia, and Mincy vividly remembered accompanying him to the dazzling gold-domed state capitol building downtown, passing time drawing on colored paper in legislators' offices while his father made the rounds.

My father and uncle were involved in government work, too, and they sometimes dragged me along on business downtown, leaving vague scenes in my mind of tall buildings and fancy lunches. Mincy and I loved some of the same drive-in restaurants, rafted the Chattahoochee River as teenagers, and watched the same local TV channels, where one weatherman, Guy Sharp, served fifty years as one of the most trusted voices in the city, though he was not trained in meteorology. Weather back then was low-tech. Guy wrote current temperatures with a grease pencil on a physical weather map when Mincy and I were kids.

We talked about beloved dogs in our pasts, learning to shoot a gun but not for hunting, and how tiresome we found Atlanta's unrelenting romance with the old Confederacy and its cursèd villain, General Sherman. (In Georgia, the 1960s were a decade-long Civil War centennial celebration even as civil rights protests were building.) Mincy and I both chose to leave the state for college, baffling many of our peers.

As we crossed the Elbert County line, I mentioned how much I had loved Elberton peaches as a kid. They were considered some of Georgia's finest. Mincy explained that most local growers of the variety had given up the orchard business. "Georgia now claims to be number one in blueberries," he said, "not peaches." The expense of intensive pesticide management of peaches and the unpredictability of the weather had left Elbert County without an agricultural distinction, but Elberton, the county seat, still calls itself "the granite capital of the world." Some forty-five quarries in the area continue to extract from a deep and so-far-inexhaustible vein of granite that runs well beyond the county boundaries.

Because of this thread in our conversation, Mincy suggested we take a little detour to see the controversial Georgia Guidestones, a Stonehenge-like arrangement of granite tablets that were inscribed in 1980 with a list of rules to retool human civilization after the apocalypse. The installation, positioned at the highest point in Elbert County, functions as a sundial, a calendar, and a compass. It was designed and paid for by a mysterious stranger named Mr. Christian who briefly appeared in Elberton. He bought the parcel of land, paid an extravagant price for the monument to be built to precise specifications by a local granite firm, gave the property to the local government, and then disappeared.

Some think the inscriptions promote a Satanic world order. Others object to Rule Two: "Guide reproduction wisely—improving fitness and diversity." They believe the monument promotes eugenics and fascism. Still others acknowledge there is some wisdom in the carved instructions, which are rendered in various alphabets and languages: Arabic, Babylonian cuneiform, Chinese, classical Greek, Egyptian hieroglyphics, English, Hebrew, Hindi, Russian, Sanskrit, Spanish, and Swahili.

Mincy relished the chance to show us the site and pointed out the last guideline on the tablets: "Be not a cancer on the earth. Leave room for nature—leave room for nature." Donna took pictures from every angle while a dozen more visitors drove up in recreation vehicles and cars with out-of-state license plates to puzzle over the inscriptions.

As we drove away, Mincy and I fell silent in the truck. Then I popped the question that had been nagging me for some time. "What is the importance of Michaux's sumac? I mean, it seems like a species we could just let go in favor of some others that are clearly more valuable."

Mincy gripped the wheel more firmly and leaned back against his seat, stretching out his arms. "Other than the fact that it is a small deciduous shrub providing browse for some animals, the *Rhus michauxii* doesn't have a great story that's tied to a pollinator or some other juicy natural history. I'm attracted to it because of its rarity and its cute little fuzzy limbs," he said, flashing a big smile. "It can survive a long time completely underground, as we have learned." He paused.

Then, suddenly more serious, he said, "I believe in the intrinsic value of biodiversity. We must value those unique genetic resources that have fought their way across millennia to survive today. My work is not just about plants that have some obvious value to humans. Our ecosystems are intricate webs. If we keep pulling strands out of the web, you never know which strand will make the whole web collapse."

When Mincy pulled up to the gated preserve a few miles out of town, Nick Holbrooks was already there waiting for us. Nick is a north Georgia mountain man, originally from the town of Toccoa, famous for its picturesque waterfall. The town is also twelve miles south of the dramatic Tallulah Gorge and Falls: a two-mile-long, quartzite canyon that was a Victorian-era tourist destination. I remember seeing pictures of my grandparents there as a young married couple. Tallulah Bankhead's paternal grandmother and the bawdy actress herself were named after the town of Tallulah Falls, and the natural landmark is also the site where the Great Wallenda strung up a cable and walked across the gorge in 1970. *Atlanta Magazine* described the stunt: "While Governor Lester Maddox and 30,000 gawkers looked on, Wallenda slowly stepped along a tightrope spanning approximately 1,000 feet across and 750 feet above the ground. After 18 minutes (and two handstands, one in honor of soldiers in Vietnam and another for laughs), the 65-year-old hopped off the platform on the other side, where his wife handed him a martini."

That story was the kind of national news that we loved in Georgia when I was a kid. I asked Nick if he remembered the publicity stunt designed to rekindle interest in the natural features of his part of the world. "Well, I wasn't alive in 1970," he told me. "But I do remember as a child when we would visit the gorge, we made a point to see the rigging where the tight rope was attached and marveled at the distance both vertically and horizontally. That was just the kind of lore we thrived on in our small corner of the world. Between that and Paul Anderson, we thought the universe spun around northeast Georgia." (An Olympic gold medalist in 1956, Toccoa-born Paul Anderson is still revered as "the world's strongest man," with several unbroken records for power lifting. A park in Toccoa bears his name.)

For his part, Nick left home to study wildlife and fisheries biology at Clemson University. He now serves as a wildlife technician with the GDNR Wildlife Resources Division and is the area manager for preserves across four Georgia counties. He is a certified burn boss and is often called in to identify and rescue wildlife in the region. He explained that we'd be hiking into the preserve along a fire break to see the bluff where the sumac were flourishing.

"You don't want the fire break road to be good enough for people to drive in here," Nick said, his eyes shaded by a ball cap. His silver mustache and beard lifted slightly as he smiled. His Georgia mountain accent was the real deal.

As Donna, Morgan, and I followed up the muddy, pitched road, Mincy and Nick traded stories about occasionally coming upon patrons with rifles and shotguns in wildlife management areas that are open for hunting and fishing like the one we were entering. For a decade, in addition to his technician duties as a wildlife agent, Nick said he was also "a badge-wearing, gun-toting, law-enforcing agent of the state."

Now, serving solely as naturalists, neither man carries a weapon, and altercations involving people with weapons are few. Both Mincy and Nick in their respective roles have run into unexpected hunters or hikers who pop out of the woods during a prescribed fire. "You just never know what you are going to find," Nick said. "Lord knows I've found the odd make-shift lab or liquor still in my day, with more anticipated."

The two also talked about the capricious nature of financial support for their work. "The lion's share of the money we receive is from a federal ex-cise tax generated by the Pittman-Robertson Act, also known as the Wild-life Restoration Act," Nick said. "It was signed into law during FDR's ad-ministration. It's a tax on firearms and ammunition. Further amendments have included other sporting goods into that excise tax. So when you've been told that sportsmen are the largest contributors to wildlife conserva-tion, that is generally how the case is made."

Paradoxically, over the past forty years, these revenues have tended to skyrocket during Democratic administrations, when people were buying guns out of fear that sales might be severely restricted. During Republican administrations, buyers seemed to feel the need to buy guns less urgently, so revenues for conservation activities tended to fall. During the Trump administration, though, extreme political polarization caused gun sales to accelerate, breaking the pattern.

"And we have to squabble for the scraps from license tag sales, grants, and fundraisers to meet our budget for plants," Mincy said.

At the federal level, more than half of the total number of endangered species on the national list are plants, but they receive only 5 percent of the funding, Mincy said. He explained that the historic exclusion of plants from the legal definition of "wildlife" dates back to English Common Law. The archaic law commanded that anything that moved belonged to the king, while feudal lords owned anything that was rooted in place. Some-how this tradition carried over into US law. Plants in the United States are protected only when they are on public or government property, when federal funds are granted to private landowners who have rare species, or when endangered plants are involved in interstate commerce.

"So if a private landowner was accepting federal money to do some-thing on their property (say, through a USDA Farm Bill program), and the activity impacted a federally listed plant species, then they would be subject to federal jurisdiction of the Endangered Species Act," Mincy explained.

Since 1995, the disproportionate application of government resources to animals rather than plants has been addressed in part by the establishment of the Georgia Plant Conservation Alliance (GPCA), the first statewide plant conservation alliance in the nation. Remarkably, Georgia's definition of wildlife included plants as early as the 1970s. Then, in the 1980s, the late Tom Patrick was named the state botanist of Georgia and began his work with GDNR as the pioneering architect of the Georgia Natural Heritage Program. Georgia is still way ahead of many others. Only sixteen states in the country have included plants among the Species of Greatest Conservation Need in their most recent Wildlife Action Plans, which are revised every ten years.

GPCA's collaborative methods for safeguarding plants are also now the gold standard for species protection and have been adopted by other states in the region. When Mincy landed his position with GDNR, working for Tom Patrick, he soon met Jennifer Ceska, the energetic founding coordinator of GPCA. "He got it immediately," Jennifer told me. "It seemed like GPCA turned on a light in him. Mincy saw the benefits of leveraging resources across conservation organizations, and he has been great at recruiting partners to GPCA." Mincy seized on the group's core values and guidelines for "safeguarding" and has been promoting them ever since.

Jennifer, who came to the University of Georgia for graduate school, was also mentored by Tom Patrick, who encouraged collaboration across the state and across disciplines. "He always stressed that recovery of natural systems requires plants, not just animals," she explained.

When we crested the hill, the flat bluff stretching out before us, my eye was drawn to the colorful patterns of charred bark on scattered pines. Some of the trees had also been cut deeply around the circumference of the trunk about three feet above the tree base. "Girdling," as it is called, kills a tree in a year or so by breaking the photosynthesis connection between the foliage up high and the roots below. First it reduces the shade the tree produces and then the trunk slowly rots, to the advantage of woodpeckers and other birds that use tree cavities for nesting. Eventually the girdled pines fall. Prescribed fire on this site also removed maples, poplars, and sweet gums, leaving only the fire-tolerant species, Mincy said. "Sweet gums," he added, "are bad to spread roots and starve other plants of water."

Just as fire was helpful to the pitcher plant populations in Alabama, the

Left: Mincy Moffett's boot provides a measure of scale for Donna Campbell's photo of the fuzzy stems of Michaux's sumac that emerged after a controlled burn in Elbert County, Georgia. Right: A male specimen of Michaux's sumac at the same site shows off its summer glory with flowers rising from the center of the plant in a conical cluster. In fall, the central cluster turns scarlet. Used by permission of the photographer, Alan Cressler.

prescribed fires on this semiforested bluff where the male sumac plants had been barely holding on have resulted in a profound regeneration. At the low point, there had been as few as two stems above ground with the overstory closing in. In 2014, after three burns over nine years, Mincy and his colleagues counted close to eight hundred stems of the dwarf sumac above ground. "Two months after that last burn," Mincy said, "it was very easy to count the new stems coming up. They were everywhere!"

Representatives from ABG systematically sampled the genetics of the new plants and found ten genotypes in the population. Considering that the population had dwindled to two stems at one time, this meant that at least eight of the genotypes had been lying dormant beneath the ground. "They were released by fire," Mincy said. "Now we take a census every year to examine the effects of the fire management." He smiled. "The mat of genetically diverse rhizomes that was dormant and had persisted underground just exploded when the habitat was right. This project increased my love for plants tenfold," he said, almost bouncing in his boots. After two more prescribed burns, in 2018 and after we visited in 2020, Mincy and colleagues counted more than 1,500 stems on the bluff. "The male and female patches have grown together," Mincy told me later by email, "and are no longer possible to separate by sex!"

The stems of the sumac, which had not yet begun to sprout leaves when we saw them in the February chill, were indeed furry, much like the velvet that emerges on the new antlers of deer in spring and early summer. Mincy pointed out one after another of the knee-high plants across the flat bluff. He said the charred loblolly pines were not natural to this habitat. Rather, the redbud, chalk maples, wafer ash, and coralberry were the norm and also indicated a circum-neutral soil in terms of acidity or pH. "Now that we know this circum-neutral soil is where the sumac prospers, we can also create a near-neutral pH in the *ex situ* nursery beds where we are cultivating more stems. We can amend classic, worn-out Georgia clay with river sand, dolomitic lime, and organic matter."

The sumac bluff had also been among the sites used to cultivate the grassland-loving Georgia aster (*Symphyotrichum georgianum*), which bears striking bright purple petals, white to lavender centers, and purple-tipped stamens that produce white pollen. The Georgia aster is another rare and threatened plant that has for now avoided the endangered list through a novel conservation agreement, called a Candidate Conservation Agreement, entered into by Alabama, Georgia, North Carolina, and South Carolina. In Georgia alone, the handsome aster is being managed on ten state lands, at a half-dozen sites each in the Chattahoochee National Forest and the Chattahoochee River National Recreation Area, and along several major power lines in the state.

Mincy said this site would also soon be used to restore Carolina trefoil (*Acmispon helleri*), through direct sowing. The perennial herb, which is a member of the pea family, has only two remaining populations in Georgia. They are on private land in Elbert County, so Mincy's team hopes to cultivate another population in the county. Also called Heller's bird's-foot trefoil, the species was identified by Heller and Small on the same day they found the Yadkin River goldenrod, described in chapter 1. Heller had first come across the plant in North Carolina the year before, and members of the Torrey Botanical Club named it in his honor. Specimens of the trefoil had been collected even earlier by Lewis David de Schweinitz, an eighteenth-century Moravian botanist, whom we will meet in the next chapter.

As Morgan dryly put it that morning on the bluff, "It's not so good to start caring about something when it becomes rare." These proactive methods that Mincy's team are using to safeguard plants were developed over the years by the GPCA, which has grown to a network of more than fifty members: Georgia universities, botanical gardens, zoos, state and federal agencies, utility companies, conservation organizations, and private companies. All of these entities work together to provide ecological land management, native plant conservation, and the protection of rare and endangered plants through proven methods. The southeastern United States supports 33 percent of the total number of plant species in the nation on just 17 percent of the land mass, according to the GPCA's website.

Safeguarding requires preserving plant materials *ex situ* (i.e., building collections of plants, banking seeds, and storing germplasm to protect genetic materials in storage facilities such as botanical gardens), but it also involves working with plants *in situ* (in nature, such as on the river bluff) by increasing the size of a plant's population. A larger population is likely to improve both diversity and viability. Safeguarding can also involve reintroducing plants in a habitat where they once lived or in a site within the plants' natural range. As a last resort, safeguarding may take the form of controlled introduction of plants where they have never been or are historically unknown, but this is the least desirable method, as we saw in chapter 2 in the case of the Florida Torreya tree—the first species that GPCA tackled at its formation in 1995.

"You have to be willing to experiment and test various propagation techniques," Mincy said. "There was a time when we thought these endangered plants were so precious, we'd just have to preserve them in gardens or in a greenhouse. And burning was not always understood or favored as a method of regeneration. But what's the point? We want the species to have ecological integrity, not be museum pieces. We are now diligently learning what these plants need. We've learned that burning works. We've also learned how to plant along moisture and light gradients—putting specimens down in the muck at the bottom of a hill and on up at different elevations on the slope all the way to the top where the soil is much drier or planting in the shade and then in increasingly sunnier spots. We have to determine what works best for the plant." This process is what GPCA's Jennifer Ceska calls "intelligent tinkering."

GPCA operates fluidly, without dues or parliamentary procedures. The work has been project-driven from the start, with only three meetings per year. "Our mission is to keep critically endangered species from winking out and to keep common species common," Jennifer explained. "Our first and best role is to protect specimens in the wild."

The GPCA also continues to enlist nurseries and private landowners to help in the rescue and resurgence of plant populations such as the Michaux's sumac. The GPCA won four national awards in recent years for excellence in the field of plant conservation. In 2013, the American Public Garden Association awarded the State Botanical Garden of Georgia the Program of Excellence Award for the creation and work of GPCA. Then, in 2018, the Association of Fish and Wildlife Agencies—a nonprofit that advocates for its member organizations across North America—recognized GPCA with a rare Award of Special Recognition, the first ever given to an organization that works with plants. "That was big," Mincy said. They also won two awards in 2019: one from the National Association of Environmental Professionals, and another prestigious award from the Federal Highway Authority, presented only once every two years.

Now, in consultation with GPCA, other southern states have launched their own plant conservation alliances. "We began working first with Alabama," Mincy said, "and then Jon Evans at Sewanee got the ball rolling in Tennessee." Mincy and Jennifer have gone on to consult with organizations underway in Colorado, Kentucky, Pennsylvania, and Maryland. "South Carolina has done well with their organization," Mincy said, "and Florida, which has so many different plant groups, has finally chosen Houston Snead, of the Jacksonville Zoo, to lead their organization." (Houston was a member of the Torreya tree-planting team whom we met in Quincy, Florida.) Mincy and Jennifer have also worked with representatives from Arizona, Hawaii, Indiana, Mississippi, North Carolina, Texas, and Washington to jump-start PCA formation in these states.

Momentum is building in the Southeast, and the second triennial meeting of the Southeastern Plant Conservation Alliance took place the week following our excursion in Elbert County. The conference, hosted by Emily Coffey at ABG, had presentations and focused discussions about developing a list called "Regional Species of Greatest Conservation Need."

To shake off the chill before heading back to the trucks, we walked over to the edge of the bluff to see the Broad River, which was aptly named and was raging from the recent rain. Mincy pointed upstream to a bend. "That's Anthony Shoals, one of the finest spots to see the shoals spider lily in Georgia," he said. A couple of local private outfitters provide float trips and tours of the flowers in May and June, Nick told us. The rocky outcrops

Morgan Bettcher (left), a botany technician, and Mincy Moffett (right), a restoration and recovery botanist, both of the Georgia Department of Natural Resources, stand on a bluff overlooking the Broad River near Anthony Shoals in Elbert County—an area where they work year-round to restore native species, including Michaux's sumac (Rhus michauxii) and Georgia aster (Symphyotrichum georgianum).

where the flowers would emerge in a few months were not visible. The lily bulbs were well underwater.

"The Broad River flows from here and dumps into the Savannah River," Nick said. I asked about shad, the unusual saltwater fish that swims this time of year into freshwater rivers along the East Coast to spawn.

"This area was a prime location for shad fishing this time of year before impoundments along the Savannah made it impossible for the native shad to make it upstream," Nick said. "Nowadays the shoals are a prime location for striped bass, white bass, and hybrid bass. Stripers are stocked in those reservoirs, and they follow a similar spawning run as the shad. They too are impeded by the dams in the river system. They swim upstream and do the deed, but the silt from up here and the flow of this water is still too slow, so their eggs get buried. The bass are doing their part, but it's for naught." Nick shook his head. "Some of those fish are harvested and mated in captivity in DNR's hatcheries. Then hundreds of thousands are stocked by Georgia and by neighboring states, which the fisherman love," he added.

Nick told us how a pair of eagles had come back to the same tree year after year near here until they were sickened by eating carrion that had fed on hydrilla—often called the world's worst invasive aquatic plant.

As we gazed beyond the slow-moving river to the tall pines across the way, Mincy elaborated on the science. "The hydrilla provides a submerged structure that supports a newly recognized species of blue-green algae/cyanobacteria called *Aetokthonos hydrillicola*. That translates to 'the eagle killer that lives on hydrilla.' The cyanobacteria produce toxins. Then paddling birds, especially coots, eat the hydrilla and the cyanobacteria with it. They develop the disease known as avian vacuolar myelinopathy and die. Then their carcasses are eaten by other birds—very frequently eagles on these large reservoirs. Then the eagles develop the disease and die."

It was sickening to me. I recognized an example of the threads of the web that Mincy had talked about earlier in the truck.

The next day, we met Mincy and Morgan at Beech Hollow Farm in Lexington, Georgia, twenty-five miles from Athens. Though the farm is a retail operation, managing director Pandra Williams and her staff perform occasional contract work for the GDNR and volunteer many hours to assist in GPCA safeguarding projects. Today Mincy's team would be outplanting Michaux's sumac on the farm as a means to test the viability of techniques used on the river bluff in a more controlled setting and to expand the scope of the restoration project with the species. Donna and I sat down with Pandra in the sturdy farmhouse that she and her husband had designed and built.

"Our mission," Pandra said, "is to see Georgia's native plants flourish and to make them available to the public. We work closely with GPCA and their brilliant botanists who educate us and help us have a deeper understanding of what these plants need. We have an employee enrichment program to teach our staff about the two hundred species we have in the nursery—information that may not be readily available in college courses these days. We're also talking about creating an internship program here on the farm for students."

Pandra was quick to tell us that her mother was a longtime environmental activist. In addition to running the farm and nursery, Pandra is a ceramic sculptor and installation artist who creates forms based on the architecture of plants and other natural phenomena. Pandra was proud to say that she attended the first Earth Day, in New York. She and her husband of forty years, Michael Williams, a retired photographer, bought these 120 acres that feature a creek and huge granite boulders scattered through the hardwoods. Once a pine plantation, the farm is now notable for its grove of beech trees in the center of the property, hence the name Beech Hollow Farm.

"All native plant nurseries are love projects," Pandra mused, sipping tea from a handmade mug. "They are very rarely a moneymaking venture. We're focusing on restoration, pollinators, and birds. It's a growth niche, I believe. All our plants are grown without neonicotinoids," Pandra said, referring to an agricultural insecticide that resembles nicotine in its chemical structure and has been shown to harm bee populations.

Pandra's passion for plants began early. "I fell in love with wildflowers on a cross-country trip in Canada in 1968. Somewhere between Toronto and Nova Scotia, we were staying in a little strip motel with huge pines out back. My grandparents were still sleeping when I went outside in the early morning. In a pool of sunlight right in front of me was a single pink lady-slipper. I had no urge to pick it. I was in awe, enraptured. I sat down and took in the birdsong and the morning."

Beyond the equipment barn in a fairly open patch near the highway entrance to Beech Hollow Farm, Mincy, Morgan, and the farm's grounds lead, Clair Eisele, were working the soil that Clair had earlier amended with sand, compost, and organic dolomitic lime. Morgan had already driven from Athens to the ABG Conservation Nursery in Gainesville the night before to pick up about a dozen female stems of Michaux's sumac that they had cultivated for this project. Mincy was in his usual GDNR uniform and down on all fours in the mud, knees wet, digging with his hands to root the first few stems.

Clair, a tall young woman who seemed utterly confident in her role, told us that she was raised in San Diego and had then spent seven and a half years in the air force before beginning her horticulture career working on organic farms. She had found her way to north Georgia through WWOOF (World Wide Opportunities on Organic Farms). Her first assignment was in White County in the north Georgia mountains. "I found my house here in Oglethorpe County on a Sunday drive," she said. Barehanded in the dirt, Clair explained that she prefers to work without gloves, to be more in touch with her planting projects.

Once the team had gotten thoroughly muddy and the stems were poking out of the soil at various angles, Donna asked the team to pose for photos. They clowned around the gray trucks with their tools, striking poses. Then, reluctantly, we bid our new friends goodbye.

Mincy reported some months later, by email, that he and Morgan had been back to Beech Hollow to limb the pines and take out a few sweet gum trees around Pandra's patch. The *Rhus michauxii* were coming along nicely, he said. Mincy hopes the females will spread out widely on Pandra's land, and sometime in the future, maybe around Valentine's Day, a few of the girls might be moved to join another population of males in a garden or in the wild somewhere.

8

River Cane

A family friend in her eighties came to visit me in North Carolina from Roswell, Georgia, where she has lived most of her life. We were driving down the boulevard that runs around Chapel Hill, heading to an early dinner of fresh seafood brought in that day from the coast. As many southerners are prone to do, Jean was entertaining herself by observing and commenting on the plantings she recognized in the landscape. Suddenly, she blurted out, "That's cane!"

Old-timers know it that way, simply as cane—the bamboo indigenous to the wild South. Botanists are likely to call it river cane, giant cane, or *Arundinaria gigantea*. This formidable species in the grass family once flourished in great, sweeping swaths across the region, and patches of it still grow where I live. It's unusually abundant near the North Carolina Botanical Garden.

In *Forgotten Grasslands of the South: Natural History and Conservation*, Reed Noss reminds us that broad bands of cane, or "canebrakes," are part of the South's deepest history. Noss tells of a visit he made to the Gray Fossil site, discovered in 2000 during highway construction in eastern Tennessee. There archaeologists made a stunning find: the bones of an ancient red panda along with dental evidence that the extinct mammal had fed on bamboo between 4.7 million and 4.9 million years ago, long before Indigenous people were on the scene.

The English explorer John Lawson was the first to document the presence of river cane in the South. He was fascinated by the many practical uses of the plant demonstrated by the Indigenous peoples he encountered. In 1709, Lawson set out from Charleston, South Carolina, with five Englishmen, picking up local guides who joined his party along the way. He traveled some 550 miles through the interior of the Carolinas. His band of explorers came across many Native communities—Catawba, Congaree, Esaw, Keyauwee, Occaneechi, Santee, Saponi, Sugaree, Wateree, Waxhaw. He also met the Tuscarora, who would eventually take his life in revenge for their displacement after English settlers created the towns of Bath and New Bern.

Though he was not a botanist, Lawson documented the landscape, the waters he traveled, and the flora and fauna that delighted him. He recorded in detail how Indigenous communities across the coastal plain and Piedmont made use of river cane as an integral part of their hunting, fishing, toolmaking, and death rituals. Lawson distinguished between the larger, hollow river cane, which, he wrote, grew "so large, that one Joint will hold above a pint of liquor," and a species of harder, thinner canes that seldom "grow thicker than a Man's little Finger, and are very rough." He was likely referring to the smaller species of native cane (*Arundinaria tecta*) that thrives in the wetlands of the South and is commonly called switch cane. In addition to eating boiled cane, Indigenous people pounded cane seeds into flour for bread.

Lawson witnessed how Native people laid fish out to dry on grids, or "hurdles," made of cane. He described how a stick of cane was sometimes applied to the mouth to punch out a bad tooth. In one community, Lawson described how, when a Native person died, the body was wrapped in blankets and then covered "with two or three Mats, which the Indians make of Rushes or Cane."

Half a century after Lawson's expedition, the botanist and illustrator William Bartram traveled much of the same territory. On the eastern side of the Savannah River, in what is now McCormick County, South Carolina (near where the Miccosukee gooseberry grows), Bartram forded the waterway and followed a Native American trading path to Mobile, Alabama. On the first day of that journey, he wrote: "At evening we came to camp on the banks of a beautiful creek, a branch of Great Ogeche [now the Little River], called Rocky Comfort, where we found excellent accommodations, here being pleasant grassy open plains to spread our beds upon, environed with extensive cane meadows, affording the best of food for our quadrupeds."

The sweet flavor and nourishing properties of young shoots of river cane would ultimately lead to the species' near demise. As Bartram noted, young cane served as ready feed for wild game and for horses and cattle, an asset recognized by the Indigenous peoples, who took care not to overgraze it. European settlers soon learned the benefits of cane to nourish their farm animals. Cattle who ate emerging cane gained weight and produced better butter and milk.

The presence of wild-growing cane usually indicated a high water table and good soil. In its early stages of growth, the nutritional value of the plant is at its peak. Settlers sought the rich bottomlands where cane grew, letting their animals graze the land before grubbing out the roots to till the soil for crops. The historian Mart Stewart says the loss of river cane is an underappreciated and key component of "the long tale of extraction and decline—of the eighteenth and nineteenth century South." River cane's surrender to field crops happened in much the same way that the region's longleaf pine forests were eventually harvested wholesale for naval stores—tar, turpentine, pitch, and rosin—and then replaced with loblolly pine plantations across Alabama, Georgia, North Carolina, and South Carolina to meet a new market demand for pulpwood and paper.

The near-wholesale extraction of river cane in the South was no small feat. According to Nathan Klaus, a senior biologist with the Georgia Department of Natural Resources, land lottery survey maps of Georgia from 1820 show that some 17,250 acres along the west side of the Flint River in Taylor and Crawford Counties were the site of a river bottom canebrake so vast and impenetrable that surveyors could find no trees on which to post their lot numbers.

The presence of a canebrake a mile wide in this area is confirmed in the correspondence of Benjamin Hawkins, a former North Carolina senator whom George Washington appointed superintendent of Indian affairs for the southeastern United States in 1795. Hawkins settled along the Flint River in Georgia the next year and took a Creek woman as his common-law wife. In his letters, he describes creeks "margined with cane," and notes that the Creek Indian farmers used cane for fencing off the potatoes and groundnuts they grew along the river. Cane originally extended as far south as the Florida Panhandle, and across all of Alabama, Mississippi, Louisiana, Arkansas, and into eastern Texas.

Following a bear-hunting trip in Louisiana in 1908, Teddy Roosevelt in *Scribner's Magazine* wrote with wonder at the impenetrability of canebrakes, describing them as "feathery, graceful canes standing in ranks, tall, slender, serried, each but a few inches from his brother, and springing to a height of fifteen or twenty feet.... Even on foot they make difficult walking unless free use is made of the heavy bush knife."

Today in some spots in the Appalachian Mountains, where historic placenames often include the word "cane," farmers swear by the value of canebrakes, which have protected their topsoil through recent hurricanes. Other folks elsewhere may be inclined to shake their heads when they see a patch of river cane, mistaking it for Asian bamboo (*Phyllostachys aurea*), and they pity the poor landowner who was foolish enough to plant it in the first place because it grows so well. But the South's native river cane creates a habitat like no other, providing refuge from predators for many species of songbirds: cardinals, evening grosbeaks, indigo buntings, hooded warblers, and water thrushes. The shy Swainson's warbler, which migrates at night and is vulnerable to collisions with the communications towers that have gone up all across the region, also depends on river cane for habitat. The Bachman's warbler, last sighted in 1988 and believed to be extinct, once thrived in southern canebrakes along with the Carolina parakeet and passenger pigeon, discussed in chapter 4. One of John James Audubon's best-known portraits is of a male wild turkey set against a background of river cane. According to the biologist Stephen Platt, historical accounts of river cane often mention the great flocks of wild turkey that once occupied stands of cane.

As many as six species of butterflies depend on river cane in the caterpillar phase, and five moth species feed on the plant. Small mammals— mice, shrews, swamp rabbits, and voles—all take shelter in canebrakes and feed on cane seed when it arrives. The canebrake rattlesnake got its name from this favored habitat, too.

Canebrakes, when they were more common, were home to bear dens. In conservation management and habitat recovery, experts today will specify cane as a high-priority plant to be safeguarded or replanted for the protection of black bear. Bison, elk, and white-tailed deer grazed on young cane and benefited from its high nutritional value.

Canebrakes served as reliable hunting spots for Native Americans, who fashioned arrows, spears, blowguns, and darts from river cane and another cane native only to the Appalachian Mountains: *Arundinaria appalachiana*, or "hill cane," as it is now known. This smaller, solid cane was recently determined to be a distinct species. Hill cane is deciduous and tends to grow along ridgetops.

In times past, canebrakes provided refuge to human beings: enslaved people on the run in the South; poor whites, sometimes called "canebrake crackers"; and those whom Mart Stewart described as "assorted hooligans who found a haven in these dense thickets of the margins."

For me, river cane conjures sweet memories tied to place and family. I fished as a child in Georgia with cane poles that my grandfather chopped and dried in his garage and strung with a hook and lead sinker on a monofilament fishing line. Bomer Henry Eubanks was born in 1893 in Cullman,

Alabama, where native cane grew far and wide on the banks of rivers. Among his people, cane poles were put to practical use to stake tomatoes, trellis beans, and catch fish.

Bomer was orphaned as a child. His mother succumbed to influenza in 1902, and his father died a few years later from a mysterious illness he contracted apparently while working in the mines around Birmingham, where rich veins of coal, dolomite, iron ore, and limestone were extracted for the production of steel. When their father died, Bomer and his three younger siblings were each farmed out to relatives to be raised. Family legend suggests that Bomer somehow drew the short straw, perhaps because he was the closest to being grown. The uncle who took him in was struggling and often fed Bomer his meals last, after the dogs. But Bomer was scrappy. He quickly trained himself to become a wily hunter, forager, and gardener to supplement his lean diet. He loved fishing with a cane pole. He also gigged bullfrogs and caught turtles that he would clean, cut up, and fry just like chicken—a practice he continued all his life, to the dismay of his squeamish and citified daughters-in-law, including my mother. Bomer was expert at growing peaches and plums and worked as a sharecropper before marrying my grandmother, who encouraged him to seek more reliable employment. In his later years, as a retired employee of the US Postal Service, Bomer instilled in his grandsons an appreciation of nature and took them hunting for deer, dove, quail, rabbit, and squirrel. For me—the last grandchild after a twelve-year gap—Bomer offered fishing lessons, and *they took,* as we say in the South.

Bomer grew a patch of cane that served as a screen between his side yard and the neighbors, "Mister and Miz Fits" (surname Fitzgerald), who were his partners in building the pond. From this thicket, Bomer would judiciously select a few stalks of just the right circumference and chop them out for fishing poles. He hung them by their narrow tips in the garage to make sure they stayed straight as they seasoned. Bomer always had several on hand for the two of us, already strung their full length with filament and equipped with hooks and bobbers. River cane makes an amazingly flexible fishing pole. The jointed segments can bend without shattering. The wood itself—a more or less hollow pipe—eventually turns from green to yellow as it dries. And once dried out, it remains strong and limber.

"You want to run the fishing line all the way from one end of the pole to the other," Bomer explained to me one day. "In case the tip breaks off from the weight of a really big fish, you'll still have the line and can keep hold of him with the piece of pole that's left."

I was never happier than on the bank of his modest pond—we grandly called it a lake—using earthworms that he'd pitchforked from the compost pile behind his toolshed for bait. I learned to wait for the red and white bobber to disappear underwater before I pulled hard on my pole to raise a glistening redbreast sunfish or opalescent bluegill, both species of bream that he had stocked years earlier.

We always ate what we caught—dusted in cornmeal and flash-fried in corn oil by my grandmother Stella before sunset on summer evenings. She served the crispy fish with more cornmeal—quail-egg-sized hushpuppies laced with chopped onion—and a delicate cabbage slaw heaped in chilled mounds on the plate. We drank sweet iced tea, sometimes garnished with mint from Stella's bed of herbs.

Suburban mansions surround that half-drained pond today. When I visited the last time a decade ago, it looked like little more than a reflecting pool. A platform and fountain had been built in the middle of the pond to produce a weak spire of water. The huge houses left little open shoreline. I suspect the bluegill and bream are long gone, as is Bomer's stand of river cane.

On the April day that Donna Campbell and I drove to Cherokee in western North Carolina, hoping to learn more about the tribe's efforts to restore river cane, the ancient spirits of the land seemed to be stirring. Wind gusts buffeted the car, quick rain showers pelted the windshield, and one rainbow after another appeared in the distance as the sun broke through the clouds behind us and shone in rays, moving like spotlights over the landscape.

A stretch of US Highway 74 leading to downtown Cherokee parallels the Tuckaseegee River, a clear stream, shallow in parts and strewn with orange and brown boulders. There we found a formidable roadside canebrake. We stopped to get a closer look.

Standing beside the guardrail of the four-lane, I positioned myself on the leeward side of the cane while Donna took pictures. According to a local weather report, the wind was gusting that day at fifty miles per hour. The older leaves from the tops of the cane ripped loose and became flying streaks of yellow, while the sturdy cane shafts rocked back and forth in slow motion like copper-green chimes. With all that bluster overhead, I did not feel the wind. When a hard gust hits the lower half of a mature canebrake, it is stopped, as in *putting on the brakes*. Suddenly the spelling of *canebrake*—rather than *break*—made sense to me. (Actually, though, the term "brake" is Middle English for "thicket.")

Cherokee, North Carolina, is the gateway to Smoky Mountains National

Park, which extends well into Tennessee. Nowadays, the center of the ancient Indian village is not the tourist carnival that it used to be when I was a child. There aren't as many shops with cheap, Asian-made hatchets, spears, and child-sized feather headdresses for sale. Historical accuracy is now the priority in tribal presentations.

The Eastern Band benefits from the prosperity of the gambling casino they opened in 1997 on the edge of town, where tourist buses arrive daily. The tribe distributes gaming revenues to improve healthcare, education, and cultural preservation. In the town center along the Oconoluftee River, the Museum of the Cherokee Indian, established in 1947 and enhanced in 1998, showcases unparalleled collections of material culture that date back centuries. Across the street, Cherokee craft artists offer their handwork in the Qualla Arts and Crafts Mutual, a cooperative gallery.

Ramona Lossie and her sister, Lucille Lossiah, are among the top basket makers at work in Cherokee today. A decade ago, tribal elders determined that the sisters were the only practicing basket makers who might be able and willing to teach the next generation and keep the craft alive. The numbers of makers had dwindled so steeply that the tribe asked the sisters to give lessons in weaving to young students in the Cherokee school system and to teach other interested adults on the reservation. Ramona and Lucille took up the challenge and began giving demonstrations at the Oconaluftee Village, a living museum of Cherokee culture open to visitors.

Lucille, now in her sixties, offers demonstrations one day a week at the village, while Ramona presents workshops all over the country. Ramona's work has been collected by the Smithsonian Institution and is represented in museums in Florida, Georgia, Illinois, and New Mexico. She also participates in competitions among Native American basket makers from tribes across the continent. Ramona is best known for her miniature baskets that carry the traditional Cherokee patterns but are made from diminutive strips of cane, a sixteenth of an inch wide and dyed with bloodroot and walnut. The delicate baskets are only two to three inches tall.

Ramona agreed to meet us at the Qualla Mutual to talk about working with river cane. Surrounded by glass cases bearing the work of generations of her elders, she was an unassuming, self-assured presence. In blue jeans and sneakers, she looked much younger than her years.

"Nothing is wasted," Ramona said when I told her I'd bought one of the miniatures on a previous visit and was amazed by the tiny patterns and the strength of the cane.

"River cane was a good babysitter and teacher when I was a child," she began. "My sister and I would sit around while my mom was weaving.

Ramona Lossie, a master weaver of river cane baskets, sits among the heritage collection of baskets and mats exhibited at the Qualla Arts and Crafts Mutual, founded in 1946 by the Eastern Band of the Cherokee Indians in North Carolina.

You learn what you see. I learned to put my things in order for weaving. I learned to count in the Cherokee language with my grandma as a part of the weaving. You are raised a certain way, and it shows in respect for your elders. My girls were raised this way, making baskets, and it shows with them, too. In our culture, if you are disciplined by your elders, it is not a bad thing. Little Native kids don't scream. That's because our elders were soft-spoken. You learned to put yourself in your weaving—no need to growl on anyone. No loud noises." Ramona smiled.

"Though we believe children are the most valuable thing there is," she went on, "we were taught that our elders need help and must eat first at every meal. Our mama ate first and told her girls to wait and be patient."

Ramona explained that her sister, Lucille, is ten years her senior and is now the elder in her family. "We are so fortunate as sisters to have each other. A lot of kids my age were lost to drugs and alcohol abuse. I am so grateful to the Creator for our family. My mom always said, 'You can say no, and you can do anything in the world.'"

Raised in the Painttown community of Cherokee, where she graduated from high school, Ramona went on to study at Western Carolina University and the University of Tennessee before living in California for a time. There she earned her pilot's license and studied for her commercial driver's license. She became a truck driver for four years, though she was often disparaged by her male coworkers for doing "a man's job." Ramona finally gave it up and came home.

"Back in the 1980s and early '90s, a lot of our elders were weavers," she said. "That's when I started doing miniatures. My sister does the bigger baskets. She says she doesn't have the patience to work with the cane scraps like I do."

Ramona told me about the family cat, Lovings, a calico who eats river cane like sweet grass. "I hear him stirring around in the cane when I am outside busting it up," she said. "He's our miracle cat. He was bit by a rattlesnake and his front side deteriorated. We put him on a pillow. The girls were crying. The doctor gave up on him, but he licked my daughter's hand and somehow, he made it. He is now eight years old and travels with my daughters to their art shows."

Monā (pronounced "Monay") and Precise' (pronounced "Precious") Lossie have earned a reputation on the powwow circuit for intricate beadwork, which they say their mother and grandmother inspired. Both young women also know how to make baskets and infuse their art with storytelling, just like their mother.

Over the years, Ramona has continued to improve her craft by sitting with the elders. "They'll only show you once," she said, her eyes shining. "I worked on my designs, watched people, learned from the people my grandmother took me to visit. I learned to dig dyes and find my materials in the woods. When you come upon a patch of cane, you never take everything. You don't get greedy. You take what you need and give the small shoots more time to grow."

The biggest challenge today in keeping the craft alive is the scarcity of suitable river cane, the local agricultural extension agent, David Cozzo, told me. "When you go into a canebrake around here, you'd be happy to get ten pieces, given the age, height, and thickness needed for weaving," he said. "The good cane is just not around here. Cane has to be at least four years old, or it is too brittle to use."

The Cherokee Preservation Foundation is a grant-making group that was launched by North Carolina governor James Hunt and the Eastern

Band when gambling was approved for the tribe. One of the foundation-supported projects is Revitalization of Traditional Cherokee Artist Resources (RTCAR), which works to preserve, protect, and teach Cherokee heritage, particularly, as the mission statement specifies, "to restore the traditional Cherokee balance between maintaining and using natural resources like river cane, white oak and clay."

Through RTCAR, Ramona, Lucille, and other Cherokee artisans have help in locating scarce river cane. Remarkably, the Eastern Band has a 1948 treaty still in force with the town of Barbourville, Kentucky, which grants the tribe access to the river cane that grows prolifically there. In exchange for this resource, the tribe sends representatives annually to the Daniel Boone Festival, where the treaty is symbolically renewed at what they call a "feast of friendship and harmony."

"We gather cane in the spring, summer, and fall," Ramona told me. "You must have the sap running. You want the cane to be limber, about the size of a table leg. If it's bigger than that, you are looking at [Asian] bamboo, not river cane."

Once the cane is harvested, the weavers quarter and peel each pole, bundle it up, and tie it off. Soaking the strips in dye can take three days to two weeks before the weaving can begin. Both of the Lossie sisters are masters of the double-weave basket made of extra-long cane splits, shaped into various forms from the bottom up. The weaver puts the shiny exterior side of the cane facing outward and first creates the basket's shape. Then she folds the extra-long cane back into the basket at the rim, creating a second, plaited section inside the basket that also has the shiny side of the cane facing outward. These complex baskets are doubly thick, and the inside mirrors the outside. The water-resistant outer surface of the cane serves to seal the basket against leakage of liquid coming in or going out. The artists may further complicate their design as they weave in different colored strands of cane at angles to create patterns, symbols, and colors on the basket's interior and exterior.

According to *Art of the Cherokee*, by Susan Power, fragments of Cherokee double-weave cane baskets dating back to 1500 AD have been found in Rhea County, Tennessee. A single-weave fragment found in Monroe County, Tennessee, is believed to be from 1450 AD. The stubborn survival of this craft is nothing short of miraculous, Power explains, given the long-running hardships that European settlers imposed on the Cherokee—particularly the forced migration of the Cherokee to Oklahoma on the Trail of Tears, beginning in 1838.

"What I see in my head comes out in my hands," Ramona said. "Like a painter, I can see the design, and when I lay my materials out, I see my hands making that design. My hands do the talking."

I studied Ramona's fingers as she talked. They were long and thin. Her wrists were narrow—a helpful attribute, I was guessing, for the complicated contortions required to blindly perform the over- and underweaving inside the mouth of a double-weave basket.

"Butternut root will give you a pretty, deep black dye," Ramona said. "Walnut will give you a dark brown, and bloodroot will turn the cane orangish." The colors of the dyed cane, still holding fast in the display cases, were woven into repeating patterns with such descriptive names as Cross on a Hill, Fishbone, Noonday Sun, Chief's Heart, Eye of the Sacred Bird, and Man in a Coffin. Ramona pointed out examples of these patterns as we walked around the room.

Photos of the artists hung above their work in the gallery exhibit. Ramona pointed to a picture of Eddie Youngbird, one of her father's people. "Lossie is my mother's family name. "Both my uncles and grandfather are weavers: Thomas Lossie Junior and Senior," she said. "They were rim makers, and they cut our wood for us. My cousin always helped me get wood for handles. He also made masks."

Ramona told me of meeting a very elderly woman who said she wove "story baskets." "Baskets tell about your family—all the designs have a story to tell. She said if she could read the designs, then she knew who she was." Ramona was silent as we looked through the glass, admiring the symmetry and patterns.

Then she said, "I always give my girls a piece of art each year. Pass it along. You work with what you've got, and you are always learning. I miss my mom most when I get stuck in the middle of making a basket. Every time I try to do a different design, I will get stuck at some point. Then, one day I'm looking at it and the basket finishes itself."

"Is weaving like being in a trance?" I asked.

"Oh, yeah. For real," Ramona answered. "If I'm weaving, I like having people talking around me while I'm working, but my girls know not to talk *to* me. They can look at my design and tell what kind of mood I'm in. We've got river cane running through our blood for who knows how long."

The small patches of river cane that persist on the margins of rivers and roadways in the South are lying in wait to reemerge and spread, should our agricultural and forestry practices change, Mart Stewart writes. Perhaps it is this latent and exploitive trait of cane that has kept it from the federal endangered species list. It is less fragile than other plants described in these pages, but a growing chorus of botanists has argued that because of its capacity to support threatened wildlife, its potential in water-quality improvement, and its carbon mitigation properties, river cane should be put to greater use.

If left to its own devices, river cane will grow thick, often from a single rhizome. The thickness of a canebrake's root system slows floodwaters and captures decomposing leaves, sticks, and other matter, which are then transformed into nutrient-rich soil. The presence of cane and its thick thatch of roots along a river helps hold soil in place, decrease erosion, and improve water quality.

In a fifteen-year experiment involving the transplantation of river cane, Nathan Klaus and his colleagues in the Georgia Department of Natural Resources witnessed a flush of growth when cane was transplanted and allowed to prosper unencumbered by non-native invasive plants in a park south of Atlanta and also in Tallulah Gorge in the north Georgia mountains, where growth has been boosted by timed burnings. Along with the flourishing cane, Nathan said, came the arrival of Kentucky warblers, white-eyed vireo, and swamp rabbits (also once known as "cane cutters" or "cane Jakes"). Swamp rabbits, Nathan told me, leave their unmistakable droppings on logs near a water source. They are secretive and quick, diving into a stream to escape detection, "but the poop always gives them away," he said, laughing.

Cane restoration projects, I found, are gaining momentum across the region. Usually the first step is removing invasive species. We visited a restoration project near Reliance, Tennessee, along Childers Creek, where it meets up with the much larger Hiawassee River. Here, Forest Service staff and volunteers removed a dense infestation of non-native autumn olives to allow the remnant cane to grow back. The next step was to fell some trees and trim others to open the overstory to more sunlight. A prescribed burn may follow, to increase the spread of the existing river cane, said Mark Healey, a Forest Service officer in Knoxville. As the Forest Service monitors the development of the existing cane in this tract, he said, it might also decide to plant "genetically appropriate river cane plugs"—cane that is close to or matches the genetics of the local stalks. The Tennessee project did not involve the use of herbicides and is one of several river cane restorations in the Cherokee National Forest. On the May morning when Donna and I visited, the trail through the forest revealed blooming trillium, spiderwort, and cardinal flowers. We walked in and out of pockets of deliciously cool air from the nearby river. The river cane, however, was sparse and spindly.

Another restoration project is developing south of the elite village of

Highlands, the town at the highest altitude in North Carolina. There, NC Highway 28 corkscrews down a steep mountainside to cross the Georgia state line and continue its descent into South Carolina, where the road flattens and crosses the Chattooga, the river made famous by the *Deliverance* novelist and poet James Dickey.

Before European settlement, the southern shore of the Chattooga River at this crossing was the site of a Cherokee village called Chattooga Town, and the road that is now NC 28 was a trading path. The Cherokee abandoned Chattooga Town in 1740, well before their forced relocation on the Trail of Tears. By late 1867, the riverside had been fashioned into a farmstead owned by William Ganaway Russell. The trading path was transformed into a wagon cut, where vacationers from the coast took a severe ride uphill to escape the heat of South Carolina's summers. The train route north from Charleston ended abruptly at Walhalla, South Carolina, some thirty miles south of Highlands, so the last leg of the journey required a horse-drawn wagon and a tough constitution. The Russell family provided lodging and homecooked meals to the pilgrims headed to Highlands, especially when the river was too perilous to cross and travelers had to wait—sometimes several days—for it to calm down. Among the guests in this era was a future president, Woodrow Wilson.

Today twenty-nine acres of cane are finding purchase along the river even as the abandoned Russell house and outbuildings decline through neglect and rugged weather. The Chattooga Conservancy launched the project in cooperation with the US Forest Service and the Eastern Band through RTCAR. The Cherokee partners hope eventually to harvest cane here for tribal artisans.

Another sovereign nation, the Poarch Creek Indians of Escambia County in southern Alabama have also focused on restoring cane. The Poarch Band is the only federally recognized tribe in the state. They are the descendants of the original Creek Nation, which occupied nearly all of Alabama and Georgia. In 2012, they received a $200,000 grant from the US Fish and Wildlife Service to restore longleaf pine and reintroduce river cane to their tribal lands. The county name, Escambia, means "river cane" in the Creek language, and the tribe's continuing work with the native bamboo has made it possible for a simultaneous renewal of basketmaking and other economic development around the town of Atmore, which is near Mobile.

Because river cane can so easily be integrated in the landscape on the margins of urban areas and alongside croplands and rivers in its native territory, and because it can grow more quickly than most trees, some scientists have called it "the poor man's carbon sink." As climate change becomes more widely acknowledged, river cane offers the individual landowner a means to participate in carbon mitigation on a small scale.

This potential made me wonder whether river cane could actually become a new commercial crop in the South. Tender shoots of edible bamboo native to other continents have long been cultivated as a human food source. Converting Asian bamboo to fabrics, flooring, furniture, musical instruments, decorative items, and cooking tools is a common practice today in Japan and China. I called David Coyle, an assistant professor in Clemson University's Department of Forestry and Environmental Conservation, to ask about the prospects for river cane as a cash crop here.

"At this point, I can't imagine that," Coyle told me. "The harvest would require painstaking manual labor, and we have no facilities to take it from the field to table or factory. But the first question really would be, Where are we going to sell it?"

Even though he was skeptical about commercial cultivation, Coyle is a firm proponent of putting river cane back where it once was. "The reason river cane grows so well here is that it belongs here," he said. "We have so many degraded river edges and wetlands in the South now, and river cane has so many structural benefits. It is one of a very few plants you can grow to secure a wetland. It's great for wildlife, especially our small birds, and it looks nice," he added. "River cane can give a nice dose of nature even in an urban area." As a follow-up, Coyle sent me a web link to the South River Watershed Alliance in Atlanta, a citizen advocacy group that has worked to improve the water quality on one of Atlanta's most challenged rivers.

Dr. Jacqueline Echols describes herself as an "enviro-community activist." Born and raised in Tuskegee, Alabama, where she also began college, Jackie eventually transferred to Clark Atlanta University and went on to earn a PhD in political science. She has worked in higher education as a teacher and administrator, but her first love is environmental advocacy. She served as chair of the Atlanta Tree Conservation Commission and as primary spokesperson for the Clean Streams Task Force in the city. Jackie is prone to preach about the power of citizen involvement in the care and protection of the natural world. She is the president of the South River Watershed Alliance.

"When I was young, Tuskegee was very rural. I spent my summers as a child following my daddy around everywhere," she said. "We fished a lot, hunted squirrels, and traipsed around in little pristine streams, getting our feet wet." In the 1990s, Jackie's activism was galvanized when federal and state authorities forced the City of Atlanta to upgrade its sewer systems and eliminate frequent water-quality violations caused by combined sewer overflows. Periods of heavy rain and flooding had caused untreated sewage to flow into streams and creeks, impacting many neighborhoods and damaging watersheds. "That's how I got started in water issues," she told me. Jackie's most recent project has focused on the development of recreational sites and the restoration of river cane along the South River in DeKalb County.

I grew up in Atlanta, but I had never heard of the South River. Jackie was not surprised.

"People don't know it," she said. "The South is one of only two urban rivers in Georgia, but the headwaters were piped and covered over a hundred years ago!"

The South River originates in a section of Atlanta that was industrialized early in the city's history and given over first to cotton-processing operations and later to fertilizer plants. "The by-products of cotton processing were disposed of in on-site lagoons," Jackie said. "Today the river first bubbles up to see daylight just north of the Hartsfield-Jackson Airport. Here the water is a bright, pale blue/white color, created by the interaction of copper, lead, zinc, and other metals that leached into the groundwater and soil."

A short distance downstream, the river is again directed underground through culvert pipes under the combined stretch of Interstates 75 and 85 before they split off in different directions south of the city. The river eventually emerges again to flow through a series of southeast and eastside neighborhoods. It then convenes with other tributaries in the Upper Ocmulgee Basin as it passes through six Georgia counties and many smaller towns before reaching Jackson Lake, a reservoir built and operated by Georgia Power Company. The river's assigned use, "Fishing," is the state's default or lowest designation and least protective of water quality. Somewhere along its downstream route, the river cleans itself up enough to meet the water quality designation that allows for fishing, but not for recreation, which requires a higher water-quality standard. Jackie's citizen organization is working to improve water quality by encouraging greater recreational use of the river. "Recreational use is the surest and fastest way to improve water quality in the river," she said, "because it forces the state's environmental regulator to enforce water pollution controls."

Below Jackson Lake dam, the river becomes the Ocmulgee River. The Ocmulgee joins with the Oconee River south of Macon and then forms the

Altamaha, the state's largest river, which flows into the Atlantic Ocean fifty miles south of Savannah.

Coming out of urban Atlanta, though, the South River has long borne a load of toxins and trouble. It still picks up rainwater polluted by pesticides, heavy metals, and petroleum products. Sewage overflows where infrastructure in Atlanta and DeKalb County have yet to be updated. Tons of trash gather in eddies and snag on sandbars. Golf courses and other commercial entities along the way withdraw water for irrigation purposes and are not required to monitor or replace what they take. Ongoing deforestation and construction in Atlanta and its outer suburban ring create an inflow of sediment that fosters erosion and changes flood patterns. All of these factors in turn damage wildlife and their food sources while also on occasion raising water temperatures high enough to cause algae bloom. This unwanted plant life cuts off oxygen, kills fish, and crowds out healthy aquatic plants.

To share the results of her organization's efforts to bring attention and remediation to the South River, Jackie suggested we meet her at Panola Mountain State Park, Georgia's first official Conservation Park. Dedicated in 1974 by then-governor Jimmy Carter, the park's namesake, Panola Mountain, is a granite outcropping, or monadnock. Like the larger and better-known Stone Mountain due north of Panola, this rock intrusion is part of a vast underground matrix of Lithonia gneiss that is also connected to nearby Arabia Mountain, another monadnock that is high enough to offer views of the skyscrapers of downtown Atlanta to the northwest. Both harbor rare plants, which grow on the rockface itself and in shallow puddles. Arabia Mountain is open to anyone who wants to hike there; however, ascending Panola Mountain requires a guide and permit because of the rare and fragile flora. All of these landmarks have been put together to form a forty-thousand-acre National Heritage Area, one of only forty-nine such entities in the United States as designated by Congress in 2006.

The countryside is especially important in Georgia's African American history. It is home to the oldest black community in DeKalb County and the site of a significant cemetery for the enslaved people who once lived there. Burial mounds of the ancient Creek Indians are nearby, making it possible for contemporary visitors to contemplate a long timeline of dramatic change on this storied landscape. Here too the South River, a reminder of the costs of industrialization and rapid urban development, winds along for some miles, flanked now on one shore by protected greenspace.

It was midafternoon on Valentine's Day when we met Jackie, bearing a gift of chocolates to thank her in advance for her time. She hopped out of her Subaru and waved energetically. She is a slim woman of enormous energy, a forceful personality clearly accustomed to rattling off data to underscore her position. A veteran of the classroom, she is both a good storyteller and a good listener. When she later confessed her age—sixty-eight—I never would have guessed it.

Jackie told us to expect a muddy hike and had thrown a bag of extra rubber boots in the back of her car in case we needed them. We had brought our own and put them on quickly. We crossed a newly constructed dam that doubles as a pedestrian bridge spanning Alexander Lake, named for a local landscape architect, Jackie said, who once owned the lake and surrounding property before it became part of the park. The spring-fed lake was newly filled after being drained to replace both upper and lower dams.

We'd driven through several upscale developments under construction in the area. This rare green expanse was already filling with late-afternoon joggers and bikers on fresh trails that fanned out from the park. We headed toward the native grasslands restoration project that Nathan Klaus of the Georgia Department of Natural Resources had launched fifteen years before and that the South River Watershed Alliance had since picked up to maintain and improve under Jackie's leadership.

Delta jets flew low and slow over us in staggered pairs, aiming for the Atlanta airport some fifteen miles west. Jackie said she would normally be at the gym this time of day. We did our best to keep up with her as she briskly traversed a hardwood and pine forest that eventually opened onto broad bottomland. Relentless rains had rendered the low-lying landscape muddy and checkered with puddles, but we followed Jackie's nimble hopscotch approach, jumping from one grassy mound to another, trying not to submerge our boots. Georgia Power, Jackie said, was a funding partner in the conservation project and owned the right-of-way, where an enormous power line buzzed over our heads as we moved through the cut.

We came upon a flock of what seemed like coastal birds picking through the puddles. "Those are killdees," Jackie said. "I always heard them called killdees as a child back in Alabama, not killdeer, which is their proper name, I believe." All at once the ring-necked birds rose in unison, calling in shrill voices before landing again a bit farther from us. Members of the plover family, they moved in quick, running steps like their counterparts on coastal beaches.

Jackie explained that the area we were in had been named "The Power of Flight," which accurately described the excellent view of low-flying commercial aircraft but also, more importantly, referred to the restored grassland area of the park where changes in bird populations are monitored. "The Audubon Society does bird banding here to collect data on feeding and migration patterns," she said, noting that the bluebird population had made a significant comeback in recent years. She pointed out nesting boxes installed on posts at intervals along one edge of the field. Part of the restoration project has involved planting pollinators for the butterfly population, she added, which is monitored in summer and fall.

Somewhere in the park, I remembered, Mincy Moffett had created a safeguarding site for Michaux's sumac. His team outplanted twenty-five females a decade ago, which now number about 750 stems, he said. Thirty males from the Lower Broad River Management Area in Elbert County where we had visited were brought down to join the females here in 2020.

Soon we saw a massive canebrake in the distance, and Jackie told us that the South River stretched along the far side of it. Her group, along with many other organizations, including Boy Scouts and AmeriCorps volunteers, had worked diligently over nearly five years to clear out volunteer trees and invasive privet so that the river cane might spread more freely. No new plugs of cane had been planted, and herbicides were used sparingly because of the proximity to the water, she said. The privet would surely come back and require ongoing maintenance, as would another invasive species that she pointed out: Japanese hops (*Humulus japonicus*). This plant, which thrives in Vietnam and other parts of Asia, was imported to North America in the 1800s for its medicinal and ornamental properties. A member of the hemp family, the vine loves a riparian habitat and, like kudzu, can grow a foot a day under optimum conditions. Japanese hops can cause human dermatitis when handled.

We marched closer to the cane. "I love to come out here in April and see the new shoots coming up," Jackie said. "I can almost imagine how it must have looked hundreds of years before."

Finding a break in the cane, we climbed a small rise, looked down, and saw slick craters scooped out of the red mud.

"Feral hogs," Jackie said. "They wallow in here. I have never seen them and don't want to. They tear things up good." It was a chilling site. I shuddered. "You would hear them coming," Jackie said, seeing the look on my face.

Georgia is third after Florida and Texas in the number of feral hogs that roam wild in the state's forests and wetlands. They have been called one of the most challenging invasive species in the South.

We stepped around the wallows to the edge of the steep drop-off to the

Jacqueline Echols, an environmental and community activist, stands in front of the canebrake she helps to maintain along the bank of the South River in Panola Mountain State Park in DeKalb County, Georgia.

river. The water was roiling, swift and muddy from days of rain. Jackie told us how, when the levels are down in summer, she organizes six-mile kayak trips for adventurous citizens. "You have to get people out on the river to see its potential and understand what's going on," she said. "We don't have an outfitter near here, so we bring in kayaks and borrow from the park as needed. The river is unbelievably beautiful and worth the extra effort to promote recreational use," she said.

The South River Watershed Alliance also hosts on-river and shoreline tire and trash collection events each year. The group is installing trash traps in some of the river's tributaries, to catch plastic bottles and Styrofoam, Jackie said, and her group is working with Coca-Cola, the Atlanta-based soda giant, in that effort.

The county had polluted the South River with abandon for sixty years and paid fines when charged, "solving nothing," as Jackie put it. While building out the area far beyond the capacity of its old and neglected sewer system, DeKalb County officials signed a federal consent decree in 2011 obligating themselves to improve the water treatment system by June 2020. Jackie already knew this deadline would not be met. Most recently, her group had filed a citizen's lawsuit to compel the Environmental Protection Agency and the state regulatory agency to enforce the consent decree and comply with the Clean Water Act. "This stuff takes years," she said, "and they know it."

I asked Jackie what it's been like as an African American woman leading in a movement historically dominated by white environmental activists. Bringing in other people of color has been a priority for her, she said, and her high-profile presence has had advantages. "I think because I stand out, my identity has opened doors. I can get in where others might not. If part of our success has come from my being a woman of color doing this work, I'll take it. This river needs all the help it can get."

As we turned away from the rushing water and headed back the way we came, a stiff breeze undulated through the river cane, mimicking waves of water. So much work undoing what people have done. The thought made me tired.

Nowadays I seem to notice river cane everywhere I go: small, upstart patches on the roadside—"squirrel tails," as one botanist described them to me. I see cane at the edges of sports fields, serving as wind and privacy screens. Sometimes a stand will be growing on the far side of a swift creek. It's sort of like how, when you learn a new word, you begin to hear that word more often in conversations or on the radio, or you notice it on the page.

Part of my plant blindness has been lifted in this research. Seeing a stand of river cane also reminds me of just how many of us it will take to do our part, to hold the ground for all the endangered species that this historic plant has harbored. Ramona Lossie's elegant miniature basket sits on my desk, reminding me how even a little bit of river cane can be useful. As Ramona said, "Nothing is wasted."

9

Schweinitz's Sunflower

Come September in the wild South, many plants begin to rust, twist, and wither. But there is always the color yellow, cheerful yellow, in the goldenrods and all manner of wild sunflowers and smaller golden asters. They brighten the darkening days until other autumn colors in the trees and shrubs fill in after frost.

When we set out to find the endangered Schweinitz's sunflower (*Helianthus schweinitzii*) in September, it was especially dry. Except for a foggy morning two weekends before, there had been virtually no precipitation in the Piedmont or mountains of North Carolina, and that was why we prepared to race toward Charlotte to catch the sunflowers in bloom. "Yes, ma'am, they are up in all their glory," the county park ranger said on the phone when Donna called to make sure they were still blooming, "but they won't last another week without rain."

Schweinitz's sunflower is a perennial that stands tall and erect, generally reaching a height of six feet, though it can grow as high as sixteen feet. Each plant has multiple flowers that bloom in September and October or until first frost. Up high, secondary stems fork off the main stem at a 45-degree angle and are topped with blossoms. The stems are often reddish purple, and the leaves are thick but narrow—oval-shaped, like the tip of a lance. The center of each flower is actually a composite of many tiny flowers that carry pollen. What I would call the petals of the flower are actually sterile yellow rays that provide a bright platform for pollinators: ants, bees, beetles, butterflies, crickets, wasps, and some species of flies. The seeds of the plant may be ingested and distributed by birds or simply fall to the ground to germinate. The plants also propagate by rhizome cloning.

Latta Plantation, our first destination northwest of Charlotte, was a cot-

and pumping gas for customers. Today the house she grew up in is gone, replaced by mansions rising around a Trump golf course. Somehow, despite all these challenges to the local ecosystems, Schweinitz's sunflower persists in a range of some sixty to ninety miles in circumference from the center of Charlotte—an area that encompasses Union County in North Carolina and York County in South Carolina.

The species, which the federal government declared endangered in 1991, seems to have been helped by its preference for poor soils unsuitable for agriculture. Alan Weakley, of the University of North Carolina at Chapel Hill, and Richard D. Houk, a professor (now emeritus) at Winthrop University, created the first recovery plan for the sunflower for the US Fish and Wildlife Service in 1994. They found that the species tends to prefer soils with a high clay content, sometimes peppered with exposed boulders and bedrock that crumbles to a fine powder.

As road building escalated around Charlotte, the NC Department of Transportation helped to create a preserve for the sunflower southeast of the city in Union County, where the construction of a dam for Cane Creek in 2005 amassed a large collection of poor soil. The sunflowers now thrive along Cane Creek, thanks to the state and to the expertise of James F. Matthews, a retired University of North Carolina–Charlotte biology professor who rescued the plants and then transplanted them on a five-acre conservation easement owned by the Catawba Land Conservancy (CLC). Visitors to Cane Creek Park can easily spot the fall sunflowers as they follow the trails for biking, hiking, and horseback riding toward the dam. Likewise, the thirteen-hundred-acre preserve at Latta features transplanted sunflowers maintained by the Charlotte/Mecklenburg Park and Recreation Department.

"It's kind of unusual for a county to take on plant conservation at this level," said Lenny Lampel, a coordinator for the department's Division of Nature Preserves and Natural Resources. Lenny is the staff member who oversees the county's herbarium collection (itself an unusual local asset). He is also curator of the county's James F. Matthews Center for Biodiversity Studies, named for the aforementioned professor, now in his eighties, who continues to be instrumental in endangered species conservation in the region.

Restoring and maintaining open grasslands, meadows, and savannas in the county's nature preserves and parks is the primary means that Lampel and colleagues have to protect Schweinitz's sunflower. Annually, in late September and early October, staff members count the stems on parklands and report their findings to the NC Natural Heritage Program and the Fish and Wildlife Service. They conduct controlled burns to reduce woody vegetation and invasive species and occasionally use chemicals and machinery to clear unwanted vegetation that threatens the sunflowers. It is an expensive and ongoing campaign.

That Lenny is not a North Carolina native was immediately apparent when I interviewed him. His Long Island accent was unmistakable, even though he has been living in the South since 2006. "To me, Charlotte seems rural," he said. "Where I came from was built out already. But here we still have a chance—in many cases our last chance—to preserve some parcels."

Lenny is a certified master naturalist, and in his current role he is responsible for the management of biological assessments and inventories, the monitoring of federal and state listed rare plant species, and the coordination of various fauna and flora studies and projects for Mecklenburg County.

"The Schweinitz sunflower is Charlotte's signature endangered species," Lenny said. "Nowhere else in the world will you find this species except in its range across North and South Carolina. Some of the largest populations fall inside Mecklenburg County." Lenny noted that several sunflower sites are publicly advertised, and others are undisclosed to protect the species.

"This is a plant that ties back to deep history," he said. "Nobody knows what this region looked like prior to European contact—the unique assemblage of plants that were accustomed here to sunny conditions. Did it look like the prairies of the Midwest? Perhaps not. It probably looked more like a savanna than a true prairie. The image of an open woodland or a prairie with a few scattered trees comes to mind, but no one knows for sure."

Lawrence S. Barden, a biology professor at the University of North Carolina–Charlotte, performed a comprehensive review of the narratives written by explorers who ventured across the region before European settlement—from the Spaniards Hernando de Soto in 1540 and Juan Pardo in 1567, to the German John Lederer in 1670, to the Englishmen John Lawson in 1701 and Mark Catesby in 1720, among others. They all described parts of the landscape around what is now Charlotte as an area of fertile prairies and open savannahs, with occasional groves of blackjack oak and post oak. Some historians assumed the lack of dense brush in the region was caused largely by frequent lightning fires that occurred during seasonal droughts. Based on his research, Barden argues that Native American fire management techniques, which improved the landscape for agriculture and game hunting, were most likely the primary factor in preserving the prairies. Once European settlers moved in and fires were not being set by residents, the land gradually became densely forested, which was bad news for the prairie plants and grazing animals. Today it's interesting to contemplate a vision of bison and elk grazing freely as they once did in those open prairies, where today NASCAR events and interstate traffic jams are the norm.

Beyond Mecklenburg County, *Helianthus schweinitzii* are scattered in disjunct locations across the Piedmont. Some populations are wild, and others have been outplanted for conservation purposes. More than eight hundred stems of the sunflower were counted in 2005 in Stokes County, on the North Carolina border with Virginia. Conservationists introduced the plants in the Sauratown Mountains in Hanging Rock State Park. The area was favored for botanizing in the early 1800s by Lewis David de Schweinitz, for whom the sunflower was named. Transplantation tends to make more vigorous plants and robust seed production, writes Steven E. Fields of the Culture and Heritage Museum in Rock Hill, South Carolina. In the case of this endangered species, as opposed to some others we have considered, scientists have little worry about genetic loss or hybridization when they are outplanted.

Another eight thousand stems were counted in 2005 in the Uwharrie Mountains, an ancient Native American stronghold in the center of the North Carolina Piedmont, now a national park that is managed by the USDA and the Forest Service. According to extensive research on the sunflower conducted over the years by botanist Moni Bates of Greensboro, North Carolina, it is likely that a site in the Uwharries or along the Yadkin River is where Schweinitz first collected the sunflower that bears his name. In 1934, Dr. Paul Otto Schallert—a physician, surgeon, and botanist who firmly and unpopularly opposed the use of tobacco—lived in Winston-Salem and taught botany classes at Salem College. He published a list of place-names he called "Schweinitz's collecting ground in North Carolina," which led Bates to this hypothesis.

The windstorms that tear through the Uwharries and down trees tend to create sunny patches where the sunflowers still emerge from a long dormancy in the ground. As Chris Matthews, another natural resources manager working for Mecklenburg Park and Recreation, told Amber Veverka of the Charlotte Urban Institute, "We get a blow-down and all of a sudden, we'll get a population of Schweinitz's sunflower. Places pop up from time to time." The North Carolina Zoo in nearby Asheboro has also protected specimen plants of Schweinitz's sunflower and Michaux's sumac and several other native endangered plant species in special gardens on the zoo's grounds.

All of these locations would seem to be good news for Schweinitz's sunflower and certainly an improvement over its status when the species was declared endangered in 1991, but several factors still keep the plant at risk. First, the populations are fragmented and do not represent adequate genetic diversity to meet the requirements to be taken off the endangered list. Populations on private property are not subject to federal protection and are therefore still threatened by commercial encroachment. State regulations in North Carolina and South Carolina are inadequate to protect the species, and resources to manage a reliable, annual census of the plants are insufficient across the range. What's more, controlled burning is not always feasible, given a site's proximity to dense neighborhoods or major roadways. Smoke can halt traffic and cause accidents.

In South Carolina the greatest density of plants occurs in eastern York County, on land that was once part of the Catawba Indian Reservation. King George III granted 144,000 acres to the tribe in 1763. Today the reservation occupies fewer than seven hundred acres, held in trust by the Catawba Nation, the only tribe in South Carolina with federal recognition.

The Waxhaw tribe, just across the state line in North Carolina and closely related to the Catawba, was still present in the Charlotte area in the mid-eighteenth century. "The area became a refuge for remnant tribes decimated by disease and cultural disintegration," Houk and Weakley wrote in their 1995 recovery plan. The two learned from historian colleagues whom they consulted that the Indigenous peoples likely grew the sunflower as a food crop. Somewhat similar to Jerusalem artichokes (also a member of the *Helianthus* family), the tubers of the sunflower are starchy and apparently quite edible. They have been compared to carrots and sweet potatoes.

Delassin George-Warren, known among his friends and family as "Roo," kindly invited us to visit the Catawba Nation. He describes himself as a researcher, activist, and queer artist, or "Two Spirit"—the traditional Native American way of acknowledging a nonbinary gender identification. Roo was born in Atlanta, grew up near the Catawba Reservation in Rock Hill, and spent time as a youth in Chattanooga, Tennessee. He studied operatic performance at Vanderbilt University and graduated in 2014. That same year he was named a John Lewis Fellow by the international human rights organization Humanity in Action, which sent him to work and study in Copenhagen. He is now a senior fellow with the organization.

Delassin "Roo" George-Warren is a citizen of the Catawba Nation who has worked to preserve the tribe's ancient language and the food sovereignty of his people. He is standing in the garden he created with tribal youth at the Catawba Cultural Center near Fort Mills, South Carolina.

After his stint abroad, Roo lived and worked in DC for some years as a performance and installation artist, lecturer, and activist. In 2017 he wrote and received a Dreamstarter grant from the organization Running Strong for Native American Youth, a group sponsored by Christian Relief Services. Roo's goal with the grant was to launch a project to revive the Catawba language. He returned to the reservation, where the last Native speaker had died in 1964. Roo taught himself the complex language even as he was teaching it to young people and their parents. He employed computer technology, toys, and games to revive interest in the language among the youngest members of the tribe.

In 2018, the National Council on American Indian Enterprise Development named Roo to its list of "40 under 40" who are making a difference. Now, as special projects coordinator for the Cultural Preservation division of tribal government, he has championed food sovereignty for the Catawba Nation. The term "food sovereignty," coined in 1996, refers to the belief that people should have the right to grow and consume healthy food that honors their cultural background and heritage. Among the traditional Catawba foods that Roo has researched is *Helianthus schweinitzii*.

Roo, who has an infectious smile and deep blue eyes, met us at the community center on the reservation. During the summer he had been teaching tribal children how to plant and care for a flourishing necklace of gardens that encircled the facility. At the main entrance to the building his group had planted an extraordinary bush: the cocona (*Solanum sessiliflorum*), a shrub native to the Andes region of South America with enormous, hairy leaves and avocado-sized fruits "that taste like a tomato," he said. The plant was definitely an attention-getter and conversation-starter.

Along a fence to the side of the building, gourds and loofas were hanging from twenty-foot vines. A few mounds of squash had yet to be harvested. Roo pointed out the mulberry trees, Catawba grapes, and blueberries set out in a nearby arbor. His students had also learned about and planted Hopi red amaranth, pineapple sage, Malabar spinach, Thai blue butterfly peas, runner beans, ground cherries, and cranberry hibiscus for tea making. "The hibiscus is a member of the okra family," Roo said. Though it was fully harvested by this time, the children had grown traditional Catawba corn: "a rematriation project," Roo explained. Mother corn, a plant that originated in Central and South America, was an essential dietary element for the Catawba.

"My grandpa made us go to his farm plot when I was a kid," Roo said. "I hated it—pulling potatoes out of the ground." But now Roo is proud of this expansive project, designed to revitalize the tribe's traditional ecological knowledge among coming generations.

We strolled through a patch of woods (hardwood and pine) to another garden patch that featured pollinator plants and a few surprises: goji berries, pignuts, sunchokes, wild buckwheat, lemongrass, and, in another area, figs. Roo had enlisted the children in building a rough log "watchtower" where they could stand guard over their crops and shoo away any encroaching animals—another bit of Catawba tradition.

When he was in college at Vanderbilt, Roo told us, he never admitted to his Indian heritage. But in his last semester, one of his music professors suggested he read Vine Deloria Jr.'s book *Custer Died for Your Sins: An Indian Manifesto*. At that point, Roo began to understand the European settlers' erasure of his heritage and the impact on his own identity, which eventually led him back to Catawba traditions, practices, sacred plants, and his people.

"If you look now," Roo said, gazing into the forest, "80 percent of the world's biodiversity exists on the landmass that was controlled by Indigenous peoples. We understood ourselves to be a part of, not separate from, nature. The Catawba lived together with the Georgia aster and Schweinitz's sunflower for centuries, with seasonal burnings to help them thrive. As we lost our land, the plants that depended on us began disappearing too, and then we lost our traditional knowledge of cultivation and caretaking."

We knew, of course, that the places we had visited across the wild South with remnant populations of rare plants had been tended carefully by earlier peoples who then had been displaced, but what Roo explained next was more disturbing. "The tubers of the sunflower have not been eaten by our people for more than two hundred years. They don't know the taste," he said. The plant's name in the Catawba language seems to have been lost too.

What's worse, precisely because *Helianthus schweinitzii* is endangered, it is more complicated for Roo to reintroduce the plant to his people on the reservation. Endangered plants were protected from digging or consumption on the federal land where we were standing.

"If we want to reintroduce the sunflower, we might have to do it on private, not tribal, land. There are so many layers of agreements with the government in our history," Roo said, with a weary outrage. "We can take tubers that are donated to us and put them in spots on the reservation that are not likely to be seen, but our people need to have a relationship with this plant, not hide it away!"

Some thirty-five hundred people are registered members of the Catawba Nation today. About fifteen hundred live on the reservation. Others who are in the US military are posted around the world, Roo said. I was surprised to learn that the predominant religious practice among tribal members is Mormonism and that another significant contingent of the tribe lives in Colorado and Utah.

With a sophisticated website and a focus on economic development, tribal leadership has made significant progress in recent years, though poverty and a high rate of suicide among youth still plague families, which is what had brought Roo George-Warren home. He says he is finding out what was lost over the years, and not just in plant knowledge.

Roo has helped to build a seed library for the tribe, and in 2020 the Nation received a grant to hire a food sovereignty coordinator to oversee the community gardens, tribal composting practices, 4-H projects, and health and wellness classes and to protect cultural, medicinal, and food plants. Roo has instituted a program to support a cohort of Food Sovereignty Fellows: high school graduates who will work for a year on the grounds, helping younger students in the garden. This workforce development program, he said, was for young people in the tribe who are not going to college, giving them transferable job skills that will provide better-than-minimum-wage work and purpose.

April Punsalan is a botanist with the South Carolina Ecological Services Field Office in Charleston, South Carolina. She and her coworkers focus on protection and recovery of threatened and endangered species. She has consulted with Roo and has heard from her colleagues that the idea of reintroducing and using Schweinitz's sunflower as a food source on the reservation—on land held in trust by the federal government— is "unprecedented." To move forward with the plant's introduction, the Catawba Nation and the Fish and Wildlife Service have to enter into a memorandum of understanding that outlines how many sunflowers the Catawba may harvest once the sunflower population reaches a sustainable level. "We don't have any direct evidence that the Nation used the tubers for food; we just have clues about this practice," April said, adding, "The Nation and the Service are excited to move forward with the project."

In 2019, the South Carolina Department of Transportation transplanted a batch of Schweinitz's sunflowers on reservation land when the agency replaced a nearby bridge over Burgis Creek. The tribe's success in managing the sunflower, whether on the reservation or on private land, will take some time yet to be realized.

Lewis David de Schweinitz, born in Bethlehem, Pennsylvania, in 1780, is honored in the annals of North American botany for his "discovery" of *Helianthus schweinitzii*, even though the Catawba tribe's familiarity with and use of the plant far preceded his collection of it in the early nineteenth century. He bequeathed all of his specimens—some fourteen hundred species preserved in his personal herbarium and collected primarily in North Carolina and Pennsylvania—to the Academy of Natural Sciences in Philadelphia. At the time of his death in 1834, the herbarium was considered the nation's largest private collection of plants and fungi. In 1842 John Torrey and Asa Gray named the sunflower *Helianthus schweinitzii*, based on a specimen Schweinitz had collected "near Salem, NC" and other specimens collected in Mecklenburg County, North Carolina, by the botanist M. A. Curtis.

Schweinitz, however, is best known as "the father of North American mycology." He was obsessed with all manner of mushrooms and other fungi growing in the moist forests of the eastern United States, where he could be found botanizing whenever he could get time away from his clerical duties with the Moravian Church. A first-generation American, born to distinguished German parents, Schweinitz reportedly developed a precocious fascination with lichen a year before he began his formal schooling at the Moravian Boys School in Nazareth, Pennsylvania, at the age of seven.

Lewis David would follow the family line, enrolling at the seminary at Niesky, in Saxony, when his father was called back to Germany in 1798. Young Schweinitz read Greek, composed in Latin, and spoke French, German, and English. He dabbled in fiction and poetry, enjoyed tobacco and a good bottle of wine, and was reported to be a witty raconteur and sociable companion.

Schweinitz's most important association in seminary was with Johannes Baptista von Albertini—a generous professor, botanist, and clergyman at the institution. In 1805 Albertini and Schweinitz published a four-hundred-page illustrated compilation of the fungi they had studied over the years. Schweinitz, by this time, had refined his talents as a watercolorist and botanical illustrator even as he progressed from student to teacher to preacher in the seminary.

The church soon called him to serve two successive Moravian settlements elsewhere in Germany, and then, in 1812, to serve as general agent of the church in the southern United States. The year is notable in that Schweinitz and his bride, whom he had just wed before leaving Germany, had to travel a circuitous route back to the States—through Denmark and Sweden—because of Napoleon's raging war.

Schweinitz and the woman he married, Louiza Amelia Le Doux, a Prussian citizen of French ancestry, survived and settled into their new life in Salem, North Carolina, in what was then Stokes County, on the border with Virginia. According to several sources, Schweinitz had plenty of time in his new role to botanize during his comings and goings across the state and beyond. He was especially taken with the Sauratown Mountains, north of Salem, and enjoyed visiting a waterfall that today is inside the boundaries of Hanging Rock State Park, one of the spots where the sunflower has been transplanted for conservation.

In Schweinitz's time the waterfall was home to an unusual collection of aquatic lichen, and for many years it carried his name. Schweinitz Cascades or Schweinitz Falls is now known simply as "the Cascades." A state historical marker honoring the botanist was placed nearby in 1979.

On his jaunts into the wilds of early nineteenth-century North Carolina, Schweinitz encountered and described many vascular plants in addition to the fungi, including the first sighting of the sunflower and of wild cannabis in the United States. The latter plant was afflicted with a fungus, which probably interested Schweinitz more than the host species.

Five years into his North Carolina position, Schweinitz received word of his receipt of an honorary doctorate from the University of Kiel in Germany, possibly the first ever awarded to an American. Following this achievement, Schweinitz was appointed to the board of trustees of the University of North Carolina in Chapel Hill—likely a partial result of his friendship with University of North Carolina professor Elisha Mitchell, a geologist and Presbyterian minister who surveyed the state's Black Mountain Range and for whom its tallest peak was named. More than once Mitchell made the ninety-mile trip from Chapel Hill to Salem to absorb doses of Schweinitz's botanical knowledge. At one or more points in his tenure as a trustee, Schweinitz was reportedly offered the presidency of the University of North Carolina, but declined, citing the distance to Salem from the university and his obligations to the Moravian Brethren.

During this period Schweinitz struck up a prodigious correspondence with John Torrey, who in 1819 was beginning to practice medicine in New York City (see chapter 1) and would go on to become professor of chemistry, geology, and mineralogy at West Point, Columbia University, and then Princeton.

"The published correspondence of the beginner [Torrey] and the established master [Schweinitz] provides an informative and amusing account of early nineteenth-century botany and of the efforts of both to obtain dependable identifications for their finds," according to Donald P. Rogers, a mycologist with the New York Botanical Garden and later a professor at the University of Illinois. (Schweinitz and Torrey would not meet in person until 1827.) Some years later, after Schweinitz's death, Torrey and his student/colleague Asa Gray named the *Helianthus* after Schweinitz to honor his memory. Torrey's and Schweinitz's lively letters, tinged with a faint tone of botanical competitiveness, were published in the *Memoirs of the Torrey Botanical Club* in 1921.

In the midst of his happy botanical work in North Carolina, Schweinitz received a summons that he accepted "out of obedience and against his will," as he put it. The Moravian church assigned him to become senior pastor at Bethlehem, Pennsylvania, and "Proprietor of the Church Estates in the North." He hated to leave North Carolina, but his new role would have him traveling even farther across the country, ultimately to the detriment of his health but giving him license to broaden the scope of his botanical obsessions.

After a period of rigorous travels for botanizing and preaching,

Schweinitz succumbed to consumption at the age of fifty-four. His widow remained in Pennsylvania, where Schweinitz was buried. Two of his sons became bishops in the Moravian Church. One, Emil Adolphus de Schweinitz, returned to the South and in 1848 was named principal of Salem Female Academy (today known as Salem Academy and College) in what is now Winston-Salem. His grandson, George Edmund de Schweinitz, served as president of the American Medical Association and was physician to President Woodrow Wilson. Adelaide Fries, Schweinitz's great-granddaughter, was well-known among North Carolina historians for her devotion as the longtime Moravian archivist and historian at Old Salem. She donated a remarkable collection of Schweinitz's unpublished botanical drawings and paintings to the University of North Carolina's library.

Despite poking some fun at the elevation of Schweinitz in botanical lore, Professor Rogers granted the following accolade to the Moravian Brother, putting him in the high company of Carl Linnaeus, the Swedish naturalist who in the 1700s invented the taxonomic classification system for living things that is still in use today:

> There are many opinions—perhaps the word might be appreciations—of Schweinitz recorded by his contemporaries and successors. With respect to his personal qualities there seems to be unanimity: he was peaceful, charitable, and notably good company. With respect to his stature and accomplishments as a mycologist there is equal unanimity, and opinions from the one side of the Atlantic are no different from those from the other; he is one of the small number of the early fathers of the science. I have seen more studies of the species, or the specimens, of Schweinitz than of those of any other botanist except Linnaeus.

After our encounter with the sunflowers at Latta Preserve, Donna and I drove south a ways toward Charlotte. We took McCoy Road, a two-lane that runs through one of the area's oldest working farms, in continuous cultivation with both crops and cattle since the 1770s according to Sean Bloom, a biologist and GIS director with the Catawba Land Conservancy. The original McCoy homeplace survives on the acreage, as does a solemn cemetery for the once-enslaved workers on the McCoy farm. Parcels that were sold or distributed to family members over the years have now been reassembled and protected as a conservation area called Gar Creek Nature Preserve: 353 acres named for the creek that flows into Mountain

The last seasonal blooms of Schweinitz's sunflower (Helianthus schweinitzii) *attracted a Gulf fritillary butterfly* (Agraulis vanillae) *on the historic McCoy Farm in Mecklenburg County, North Carolina.*

Lake, the main drinking water source for Charlotte. Mecklenburg Park and Recreation owns and manages the preserve, and the CLC holds it in trust.

Thirty-six of the acres have been pulled from active farming and restored as Piedmont prairie, marked with signs on the roadside that discourage visitors but celebrate the prairie's glory. Conservationists went to work when the Schweinitz sunflower was found growing wild on the farm in 1992.

It was not necessary to bring in additional sunflowers for the prairie project, only to improve the habitat that the plants had chosen, Sean explained. "A lot of our rare and endangered plants are found in high-quality sites, but Schweinitz's sunflower, no matter what, it seems, wants to grow right on the side of a road somewhere," he added and laughed.

We slowed down to look more closely, and there they were in the field by the road, tall and waving in the breeze as the sun dropped. The yellow discs were receiving guests: bright orange, brown, and white butterflies gingerly sipped at blooms not yet withered. The sky was what Tar Heels call Carolina blue—intense and unsullied by a single cloud.

10

American Chaffseed

When the botanist Dwayne Estes was in sixth grade, his Tennessee history teacher, Tommy Johns, told the class an apocryphal story. Dwayne—who was quick to admit that his teacher paddled him thirteen times that year for misbehaving in class—told me he listened well on that particular day. The story was memorable.

As the pilgrims aboard the Mayflower arrived on this continent, Johns told his students, they saw a deep, dark forest just beyond the rocky coastline. It was the same view, the teacher said, for those who landed at Jamestown, and even for the explorers who came ashore at St. Augustine. As Tommy Johns told it, the Eastern American landscape was a forest so dense and continuous that a squirrel would be able to travel all the way from the Atlantic Ocean to the Mississippi River in the treetops, his paws never touching the ground.

"You can still imagine landscapes like that if you're traveling through the Great Smoky Mountains," Dwayne said. "There were big trees back then—tulip poplars and chestnuts that were ten feet in diameter—and there were some very large forests, but that myth of the squirrel is an outright lie!"

In the many talks that Dwayne now gives around the region in his position as cofounder of the Southeastern Grasslands Initiative (SGI), headquartered at Austin Peay State University in Clarksville, Tennessee, he repeats and debunks the squirrel story and then begins to describe the once-prolific grasslands of the Southeast, their intense biodiversity, and how many rare grassland plant species we are on the brink of losing because so much land has been given over to forests.

Grasslands, he reminds us, support threatened and critically endangered pollinators and other animals such as quail, deer, turkey, and elk.

The environmental movement's tendency to focus on protecting forests has come at a great price, Dwayne argues. Today some 95 percent of the historic grasslands in the wild South have been eradicated—a dire situation that SGI is trying to reverse. The organization has been expanding rapidly in prominence and philanthropic partners since Dwayne and his cofounder, the eminent Arkansas botanist Theo Witsell, launched the effort in 2017.

"I suppose I might have learned about grasslands if I had been a student in some classroom in Kansas or South Dakota, but I was not," Dwayne says. "I was born and raised in southern middle Tennessee, in an area near the Alabama line. And what most people raised in the South in the last hundred years simply never learned is that grasslands, meadows, longleaf pine savannahs, dry prairies, grassy balds, and open wetlands were all a part of the landscape of the region long before the camera was invented."

Or, as Arizona State University professor emeritus Stephen J. Pyne bluntly put it in his book *Fire in America*, "The virgin forest was not encountered in the sixteenth and seventeenth centuries; it was invented in the late eighteenth and early nineteenth centuries." Human beings literally created many of our forests through neglect and destruction of the native grasslands. Restoring these biodiverse remnants is an urgent task. In another twenty-five years, Dwayne figures, it will be too late.

Estes has a ready arsenal of memorized quotations that pepper his presentations. Among his favorite witnesses to the early dominance of southeastern grasslands is Elder Reuben Ross, a Baptist minister. Born in 1776 and buried in Clarksville, Tennessee, where Dwayne now lives and teaches, Ross roamed the region and fancifully described in 1812 his view of the Pennyroyal Plain Prairie of southern Kentucky and northern Tennessee: "It would be difficult to imagine anything more beautiful. Far as the eye could reach, they seemed one vast deep-green meadow, adorned with countless numbers of bright flowers springing up in all directions. ... Only a few clumps of trees and now and then a solitary post oak were to be seen.... Here I first saw the prairie bird, or barren-hen.... Here the wild strawberries grew in such profusion as to stain the horses hoofs a deep red color."

In 2020 Austin Peay and SGI announced a new partnership with Google, Roundstone Native Seed, and Quail Forever to plant fifty acres of tallgrass prairie on the southern edge of what was once the Pennyroyal Plain Prairie, also known in 1800 as the "Big Barrens." The restoration project, located near Google's Clarksville data center, will be planted with a diverse mix of some sixty species of regional genotype seeds. Restoring habitats where native plants still thrive in small numbers is central to SGI's mission, as is taking inventory of prairie remnants in the Southeast.

Though the pervasive existence of the grasslands was blotted out of our collective memory for more than a century, many of the species that occupied the landscape are still with us, albeit mostly as buried treasure. It is this treasure trove of native plants in seed banks underground that Dwayne Estes and his colleagues are promoting to philanthropists, corporations, and the public to appreciate and preserve. In addition, SGI's chief botanist, Alan Weakley of the University of North Carolina, is working to bring attention to more than seventy-five southeastern grassland plant species that are so new to science they have yet to be named.

Dwayne's Tennessee twang flavors his speech, even when he reels off the Latin names of every plant in sight. He drives a black Toyota pickup truck with a camper top, and when he resorts to a coat and tie for the more formal presentations that he gives at garden clubs and academic conferences, he is not exactly in his element. Dwayne is six foot three with curly black hair and a neatly trimmed beard. Though he serves as professor of biology, director of the Austin Peay Herbarium, and principal investigator for the university's Center of Excellence for Field Biology, he has not a shred of academic pretension. His humility is earnest and his passion for the natural world is infectious.

Dwayne was raised by a single mother with limited resources. For many years he suffered from shyness in group situations. "I never felt secure in my ability to approach people," he said. As a boy, he spent many hours alone, exploring the fencerows and thickets of the countryside, and would sometimes stray a mile or more from home when he was only seven years old. School was also a challenge. "We were not a reading family," he said. "The only book in my house was a 1968 Funk and Wagnall's encyclopedia."

When he reached the eighth grade, Dwayne decided to transform himself.

"I wanted to be popular. I wanted to have a girlfriend. I wanted to make straight As," he said. Dwayne launched his self-improvement campaign in the Ridgeforth Middle School Library, where he vowed to himself to read every book on the shelves before he graduated.

"The first book I opened, I looked through a few pages and put it back on the shelf and said, 'I'll get to you later.'" Then Dwayne pulled out the next book, William Carey Grimm's *Indian Harvests*, an illustrated volume published in 1974, describing forty North American plants that had been important to the Native American diet. A fascination with natural history overtook him that day as he read about making sumac tea and picking

off the cleared land to small farmers. When the Depression hit, timber and coal interests pulled out. Gradually this land was acquired in parcels by Tennessee and given over to the state wildlife management agency to be used for public hunting, fishing, and recreation.

"Wealthy folks back in the day retreated to the Cumberland Plateau," Dwayne said as we bumped along the first few miles of gravel in the refuge. "The altitude and the forested landscape keep the temperature down. People in the twentieth century came up here for cool air in the heat of summers, and many still think the forests are what should be here. The plateau has long been promoted as a forested landscape, and it has continued getting denser and thicker. By the 1980s so much of the plateau was forested or given over to pine plantations that species needing an open landscape really took a nosedive," Dwayne said. "One of the main casualties in the 1980s was the red-cockaded woodpecker." The last woodpecker specimens from Catoosa were taken to Fort Benning, Georgia, in 1990.

Dwayne drove over a rise, and at the crest, the view opened up. The area ahead had obviously been burned on both sides of the gravel road, and here the trees thinned out. We had arrived at the thousand-acre grassland restoration that had begun as a timber salvage operation.

"The southern pine beetle was coming this way, and they knew it would ravage this forest that was then choked up with old-growth hardwoods and pines," Dwayne explained. "The Tennessee Wildlife Management Agency decided to clear the valuable timber—tens of thousands of board feet of old-growth pine—before the pine beetles got to them."

Dwayne and his graduate school mentor from the University of Tennessee, Dr. Eugene Wofford, visited the area shortly after the cut, approaching the acreage from the hemlock gorge on the opposite side of the open area where we stopped to take in the view. The two botanists were furious. "It looked like a brutal clear-cut, a scar on the landscape. Dr. Wofford cursed and I went along with him," Dwayne said. "We went on being pissed off about the obliterated forest until years later, when I met Clarence Coffey."

Coffey, a biologist and expert on shortleaf pine (*Pinus echinata*), is now retired from the Tennessee Wildlife Resources Agency. He had been assembling historical data on the acreage for years and predicted a tremendous resurgence of wildflowers and grasses along with shortleaf pine, as long as the landscape was burned after the trees were cut to protect against the pine beetle.

"And he was exactly right," Dwayne said, "though cutting that timber was a tremendous leap of faith at the time." Once Coffey connected the dots for Dwayne, he became a big fan of Coffey's historical research techniques.

"There are a lot of places in the coastal plain of the South that you could go to now and you would never know there were longleaf pine savannahs there at one time," Dwayne said. "They were either logged out or converted to other land uses. The same is true here. There are many academicians now who question whether the shortleaf pine was ever a major component of the Cumberland Plateau. I would suggest that people have forgotten to look at the historical record."

Like Coffey, Dwayne now looks beyond scientific journal articles and naturalist narratives to study real estate plat maps, genealogical records, government documents, old newspapers, and other archival materials to find evidence of the previous existence of grasslands, wildflowers, and certain tree species in Tennessee.

As Dwayne put it, "Coffey found that shortleaf pine had been abundant up here. Historically, it supported a whole industry and many livelihoods. It supported species like red-cockaded woodpecker and American chaffseed, and then both vanished. In some places today, shortleaf is so rare you'd almost think its absence was purposeful—that people were trying to get rid of it. But you're going to be delighted to see that it's making a total natural comeback up here, all on its own."

We stopped and got out of the car as Dwayne pointed to a grand specimen—a forty-year-old shortleaf pine. "The Tennessee Wildlife Resource Agency deserves full credit for this project," Dwayne said, noting that across many southern states, the work of wildlife officials has not always been seen as science based. Wildlife rangers have been pejoratively nicknamed the "hook and bullet guys." But the comeback at Catoosa has been so impressive that some officials from other agencies involved in conservation suspected that it was created: that the plants were seeded in rather than the product of a natural process.

"I wish we had the capability to reseed with these results!" Dwayne said, his voice rising. He was starting to get exercised. He moved into preacher mode. "This savannah restoration should be a model for hundreds of thousands of acres in the South. Three hundred species of plants returned, including rare species that have come back." He stopped. And as if on cue, a bobwhite quail called from the brush beyond the shortleaf pine that Dwayne was leaning against.

The call of a bobwhite—first with a short note and then a longer lilting note—comes out as two quick syllables, as if the bird is calling its name, "bob white," or sometimes "bob bob white." It was the first bird call I recognized as a child. My grandfather, who taught me to tell time and to whistle before I was in kindergarten, also acquainted me with the songs of the birds in his yard. These days, where I live, it is rare to hear a bobwhite quail. As I later learned from shortleafpine.net, a website for conservators of the species of which Clarence Coffey is a prominent member, the grasslands and the shortleaf pine provide an important food source for small mammals and birds, especially northern bobwhite quail (*Colinus virginianus*). The shortleaf also offers roosting habitat for bats and roosting and nesting habitat for woodpeckers. The northern bobwhite and the eastern meadowlark have declined by 75 percent over the past forty years, and some predictions suggest that they will decline by half again in the next ten to twenty-five years.

We listened as the bird called again, then suddenly a bright orange and brown butterfly floated by and Dwayne was off to chase it with his iPhone. "I think it's a Diana fritillary!" he called out over his shoulder. We quickly looked it up on the Seek app. The Diana is the state butterfly of Arkansas and can be found in several counties in South Carolina, and in spots along the Appalachian mountain range. Dwayne was thrilled when it landed on a black-eyed Susan and he got a picture, and Donna got a picture of Dwayne taking a picture. It was fun to botanize with such an energetic consumer of nature.

As we got back in the van, Dwayne talked about his vision for the Cumberland Plateau and the need for the kind of restoration that is already going on in Georgia, North Carolina, and South Carolina in the form of quail plantations and savannah grasslands that can support rare and native species. "I'd like to see more thousand-acre estates like this one returned to grasslands here," he said. SGI already shares four staff positions with the organization Quail Forever, which is an important nonprofit partner in Tennessee, but there is much yet to be done.

"Based on the forensic clues that I look at as an ecologist," Dwayne said, "somewhere between 40 to 60 percent of the Cumberland Plateau is flat to rolling and would be eligible for grasslands restoration, but many in the scientific community won't look any farther back than 1870 to see what preceded these dense forests."

New technology called lidar mapping can generate precise, three-dimensional information about the surface characteristics of a landscape even when there is a dense tree canopy. Lidar, which stands for "light detection and ranging," uses a pulsing laser to measure topography variations down to a few millimeters. Helicopters and airplanes are used to gather lidar data over a discrete area. According to Dwayne, the technology has been used to find Mayan roads and shrines in the Yucatán and pre–Civil War tar pits in North Carolina's Croatan National Forest. It has also proved that the Cumberland Plateau was a prime region for the production of pine products during the same era, but for some reason that period in the economic history of Tennessee is not as well-known as that of North Carolina, he said.

With a grant from the Fish and Wildlife Service, Steve Simon, an independent plant ecologist out of Asheville, North Carolina, visited two thousand sites on the Cumberland Plateau to set data points to be used for mapping the geology, soil, elevation, and vegetation types—especially fire-adapted plants—to help determine future conservation goals. The grid drills down on thirty attributes of the landscape, which were examined in six-meter squares. The results show that some 40 percent of the Cumberland Plateau would thrive as pine savannah.

Thinning out the forest to restore grasslands can be highly controversial, but as Dwayne puts it, "the loss of southern grasslands is the greatest threat to terrestrial biodiversity in North America. Yet it is not even on the radar of most conservation funders and many conservation organizations."

As we drove on, admiring the restored grassland, Dwayne continued, "We have done a great job conserving rocks and ice in our mountains, but we have not given the species that need open landscape what they need— neither plants nor animals. *Schwalbea* is emblematic of hundreds of plant species. In fact—and this is the greatest warning to us about the plateau grasslands—there were forty rare species of plants in the acidic wet savannahs on the plateau of Tennessee alone. Of those forty species, we've lost half of them in the past twenty-five years." Dwayne shook his head.

The mismanagement of upland wetlands has been ongoing, but the destruction of critical habitat began when mechanized machinery, such as bulldozers, were introduced to the Cumberland Plateau after World War II. And now with climate change, Dwayne argues, grasslands restoration should be a priority to improve water quality, sequester carbon, and bring back native species. Bringing back the plants, he says, brings back insects, bats, and birds with them.

These goals were what Google had in mind when they came to SGI to discuss the restoration project in Clarksville near their data center. "And selective timbering—not clear-cutting—can help generate the funds to

Dwayne Estes, executive director of the Southeastern Grasslands Initiative, stands at the edge of the Devil's Breakfast Table overlooking Daddy's Creek Gorge in the Catoosa Wildlife Management Area in Morgan County, Tennessee. American chaffseed (Schwalbea americana) was last spotted in the 1930s in this region.

pay for the restoration," Dwayne said. The Band Foundation in New York has been the biggest investor in SGI from the beginning and has been a steady partner as the organization has grown.

The only remaining unplowed natural grassland in Tennessee today is May Prairie State Natural Area in Coffee County, home to more rare species than any other area in the state. It was there that Dwayne identified *Symphyotrichum estesii*, a new species of aster that now bears his name. Other species previously unknown to science were also recently identified near our next stop, in the Catoosa Wildlife Area. We soon parked and headed into the brush.

Dwayne's students were assigned to study the river scour habitat at the bottom of the gorge at Daddy's Creek, where dangerous whitewater and sixty-eight-foot-deep pools attract only the most daring kayakers. At the end of a short trail, we leaned gingerly over a rock outcropping to see the water below. The landmark where we were standing was called Devil's Breakfast Table, a rocky glade where Dwayne totted off an array of plant species surrounding us, including lowbush blueberry (*Vaccinium angustifolium*) and bastard toadflax (*Comandra umbellate*), with its elegant white flowers.

When André Michaux came through the Cumberland Plateau for the first time in 1793, he rushed through this area out of fear, Dwayne said. Besides the rugged landscape, Michaux worried about the Native Americans, who maintained a toll road here and were known to attack travelers. The famous botanist did not pause for much note-taking on the local flora and fauna. After spending the night at nearby Crab Orchard, Michaux had traveled, he wrote, "in company with 12 persons who had assembled at that place to pass through the Woods inhabited and frequented by the Savages. The tract between Crab orchard and Houlston settlement is 130 Miles wide and is called the Wilderness." After arriving at Houlston after seven days, Michaux concluded: "My horse was so tired owing to the rapidity of the journey and the bad roads across the Wilderness that I was obliged to stop after a Journey of only eleven Miles." In that journal entry Michaux had also described "a climbing fern covering an area of over six acres of ground near the road" at Cumberland Gap, but little else of a botanical nature, meaning that his estimable voice was missing from the historical record describing the elegant and complex ecosystem of grasslands in the Cumberland Plateau.

"There's a whole suite of plants endemic to this area from the scour of water on rock down there," Dwayne said, pointing down into the gorge. "There's a new clematis, one euphorbia, and a new species of big bluestem grass." His students at Austin Peay are in the process of eagerly introducing these new plants to the field of science.

As we finished our tour, Dwayne told us of plans to outplant *Schwalbea americana* in Tennessee as part of a future restoration project. The plant is the only species in this book that we never saw in our travels, either in the wild or in cultivation, because as we would learn from April Punsalan, the Fish and Wildlife Service botanist headquartered in Charleston, the reintroduction of the species can be tricky.

April's personal story was somewhat similar to Dwayne's. She was one of three children raised by a single mother in the inner city of Norfolk, Virginia. "I didn't like school, and I didn't fit in," she said. Then she took a class in horticulture with an award-winning teacher, the late Betsy Spears, at Norfolk Technical Education Center. April began reading the magazine

Birds and Blooms. She participated in herb competitions and won. Then she was elected president of her school's chapter of Future Farmers of America.

After high school April worked for a time at the Norfolk Botanical Garden. "I was always interested in plant medicine, and I came to feel like we have a responsibility to give back to plants for their gifts to us," she said. She began turning to medicine plants to cure herself whenever she came down with a cold or other ailment.

April enrolled in the University of North Carolina at Asheville for her undergraduate degree in environmental science with an emphasis in botany, taking every plant course she could find and working as an intern with the US Forest Service. She then earned a master's in science at Western Carolina University. Her first job as a botanist with the USFS took her to the Nantahala region in North Carolina. Now she works for the US Fish and Wildlife Service in South Carolina, and her husband is an outreach and special events coordinator for the City of Charleston. They are raising their daughter to appreciate the coastal plain.

As South Carolina's lead botanist for *Schwalbea*, April has been working to encourage habitat protection and management for the chaffseed. South Carolina supports the largest populations of this species and is the heart of its range in the South. "We are so fortunate to have these quail farmers who annually use fire as a tool to maintain the longleaf pine savanna habitat that is critical to the survival of quail and many other imperiled savannah species," she said. April believes strongly that because human beings shaped the landscape, we have a responsibility to repair the damage we've done. "We are the keystone species here," she says, meaning that the human population has had a disproportionate effect on many ecosystems in the region.

April let me sit in on a conference call among horticulturists and botanists up and down the East Coast, from Duke Farms in New Jersey to Tall Timbers in northern Florida. All of them were in various stages of cultivating *Schwalbea*. They talked about the timing of seed collection and propagation, whether to plant the seeds with a host and what kind of host, what types of soil mix they were experimenting with, the watering regime, the size of the pot, and when the plants might be hardy enough for outplanting.

The legendary ABG naturalist Ron Determann was mentioned. "He could root a pencil!" someone said, and they offered a recipe for Ron's special soil mix for American chaffseed: it included charcoal, peat, and sphagnum. Others were working with native soil enhanced by peat, pearl-

April Punsalan, a botanist at the US Fish and Wildlife Service headquarters in Charleston, South Carolina, coordinates with botanists along the Eastern Seaboard to conserve American chaffseed (Schwalbea americana) in the wild.

ite, and vermiculite. Some had germinated the seeds in a cold setting; one had done so in an orchid greenhouse with domes and misters. The preferred host plants were asters, useful for their dense root systems. One grower was using fish fertilizer every two weeks.

April did a masterful job keeping the discussion moving forward and finally asked the group to consider whether they were putting too much energy into cultivating plants with very mixed results at the expense of looking out for the extant populations in their states. Operating on a limited budget and meeting the recovery criteria to take *Schwalbea* off the endangered list was a numbers game, according to federal guidelines. April said her goal is first to manage the habitats carefully and to continue encouraging those who've had success in outplanting. Clemson University researchers have had the best results in South Carolina in propagating the species.

"Reintroduction of *Schwalbea* needs to be planned and well thought out, but the first priority must be to protect the populations we have," April said. The Endangered Species Act emphasizes that captive propagation should be used only when efforts to maintain and manage the species' habitat fails or has shown to be ineffective to recover the species. It's critical to use resources to manage the habitat of the species first before performing reintroductions. "Pollinators need a larger number of patches to keep the population numbers up to sustainable levels," she added.

For her master's thesis at Columbia University, a promising young botanist named Brandi Cannon focused on the conservation of *Schwalbea americana* by evaluating the genetic diversity of *Schwalbea* populations across the eastern United States and how various host plant's root systems responded to fire. Her work is cited in the US Fish and Wildlife Service's 2019 report on the species in the Southeast that April helped to prepare. Brandi noted that parasitic chaffseed is particularly important in helping to control the growth of certain grasses that, if left unchecked, can disrupt the delicate balance in a grasslands ecosystem.

The grasslands restoration and conservation discussed so far in this chapter has been restricted to relatively large tracts or preserves, but what about the value of conserving and restoring grassland species in urban settings? Annabel Renwick, curator of the Blomquist Garden of Native Plants at Duke Gardens, seeded a miniature prairie on the Duke campus and has so far conducted one burn on the parcel. Now in its fifth year, the mini-prairie habitat was too wet during the most recent burning season and the window passed, though a burn would have been preferable.

Left: Brandi Cannon photographed this specimen of American chaffseed (Schwalbea americana) during her master's thesis research at Columbia University. Right: The botanist Brandi Cannon, photographed by Annabel Lee Russell. Brandi is a middle school science curriculum coordinator at the Dwight-Englewood School in New Jersey and an organizer of Black Botanists Week, launched online in 2020 to identify and encourage people of color who appreciate the global diversity of plant life and to create and promote a safe space for them. Used by permission of the photographers.

Annabel says she's been "editing, pruning, weeding, and taking inventory of the perennials that have come back. Pulling up a lot of blackberries, too," she added.

Many of the native grasses she planted in 2015 are being outcompeted by goldenrod, ironweed, and helianthus—all aggressive species, she said. Most of the hundred original species are present, but not in the same abundance as when they were planted. Annabel found it ironic that when a North Carolina forestry expert arrived to assess the mini-prairie for burning, he declared, "It looks so natural!"

"Natural is not what the majority of people come to the largely formal Duke Gardens to see," Annabel said, laughing. "Dwayne Estes has his preserves, and I am trying to achieve something semiwild in an urban environment. This area is a contrast to the rest of Duke Gardens and represents a landscape that once covered much of the Carolina Piedmont. At Duke Gardens we welcome more than five hundred thousand visitors a year. They come to see the spring bulbs, the ducks, and, in summer, the amazing floral borders."

When guests enter the prairie, however, often by chance, it stops them in their tracks. Annabel explained, "It is not what they expected as they wandered around the corner. Often they turn around and retrace their steps, but sometimes the more inquisitive will ask about the landscape."

One year a group of children from a summer camp came to visit the garden. Watching the group begin their tour, Annabel saw that the camp supervisors found their phones more interesting than the plantings. One young boy, around twelve years old, came up to Annabel and asked, pointing to the prairie, "What is this all about?" She told me, "I was overjoyed by that simple question from an inquisitive young man."

Annabel, whom we met earlier in the Torreya tree chapter, was raised on a hill farm in northern England, where hardy breeds of sheep and cattle roam the moors, locally known as fells. These animals are "hefted" to hillsides. (The term "heft" is from Old Norse, meaning that the animals become accustomed to specific hillsides, she explained—that they have a provenance.) Early on, Annabel studied the effect of stocking densities and grazing on plant communities in various types of grasslands in the United Kingdom. Just like prairies, fells are burned to reduce the shrubby growth of heathers, and they are grazed rather than mowed.

"We know that the prairies in this country once had large grazing herbivores, such as bison, elk, and deer," Annabel said. "But how much do we know about the preferential grazing of the various animals? How did the historic populations graze? We know a cow tends to graze higher, needing longer vegetation than a sheep, which grazes close to the ground, and like a horse, a sheep can nibble, but a horse can pull the plant out by the roots."

Many questions linger about the historic herbivores that grazed in our southeastern grasslands. What were the natural stocking densities of the herbivores? Were these animals naturally hefted to certain regions, or did they roam for hundreds of miles? Did they follow the same trails? If they did, the vegetation around the trails would be different, Annabel explained.

"The trails areas would receive a greater quantity of natural fertilizer, and the soil would be richer, impacting the type of plant that would grow there. Sheep tracks can be clearly seen across moorland," Annabel said, but it is very difficult to mimic the impact of the grazing on a landscape with mechanical mowing. The height of a mower can be varied, but the recycled nutrients in animal waste are not present when there are no animals.

"Obviously, this is something I don't have to consider with the mini-prairie at Duke," Annabel said. "But in my opinion the presence of large herbivores should be considered in the management of historic grasslands."

I had to admit, this prospect was not something I'd thought much about.

For fifteen years Annabel and her predecessor at Duke, Stefan Bloodworth, offered monthly public talks called "Walk on the Wild Side." A recent presentation was on conservation and included the Piedmont prairie. Annabel reminded her audience that the population of the planet has doubled since Earth Day in 1970 and that by 2050, according to current estimates, 75 percent of the world's population will be living in urban areas.

"So how do we take care of wildlife and humans going forward?" Annabel asked the audience. "How will we manage to share our urban areas with essential plants?" She hopes that the prairie at Duke Gardens will serve as a portal for the public to see native plants in a natural setting while being able to imagine the ways in which such plants might also be brought into urban sites—in roundabouts, on roadsides, and in greenways.

Annabel has helped one company create a native prairie project on its corporate campus at nearby Research Triangle Park. The site contains more than forty species of local ecotype native plants. Annabel supplied the seeds, and company employees propagated thousands of native plants over a period of three years. She says that planting such habitats is one thing and managing them is another.

"We are still trying to understand how to manage these types of natural urban landscapes. If we want the idea to spread across corporate campuses, it is critical that the landscaping companies who work on these sites know how to take care of the plantings in ways that are very different from what they are accustomed to. It cannot just be blow, spray, and mulch," Annabel said. She believes there is a desire to see more natural landscaping.

"There are organizations in our region starting to appreciate the value of pollinator gardens and natural landscapes. Without a doubt, creating mosaics of plantings throughout the urban landscape would provide corridors or refuge for animals and insects," she added. "At Duke we are also evaluating how native plantings might perform in the medians of parking lots and other small urban spaces. These urban spaces are far removed from the natural preserves, but they are places we need to work with to encourage the growth of plants that attract not only the butterflies and bees but also humans." Annabel concluded, "Connecting people with plants and nature is critical for our well-being."

Conservation may come down to each person doing their part in their place, as we have seen among the botanists, naturalists, conservationists, and laypeople whom we have interviewed here and who are working to preserve endangered species. The eminent biologist Peter White, who served as director of the North Carolina Botanical Garden for many years, has written about the ethics of environmental sustainability in the twenty-first century—how in the United States and Europe it depends on working in patches and remnant landscapes, and restoring historic ecosystems that have been interrupted by human habitation and agriculture. White and his European colleague Anke Jentsch echoed what April Punsalan told me about her personal ethos as a botanist. These seasoned scientists wrote that "humans cannot be considered as external to the ecosystems on which biological diversity depends." Biodiversity is the key to the survival of all species.

As Annabel pointed out, the human population is growing larger and having a greater impact on ecosystems than ever before. Our presence "reduces the space available for ecosystem dynamics at the multipatch scale," White and Jentsch explained. "This has two implications: that the ability to find solutions at large is being reduced and the ethical imperative is therefore shifting to smaller scales."

Conservation biologists are working in smaller and smaller bits of landscape. Poor decisions made on a small scale that might have once been tolerated as part of a larger landscape are now reaching a threshold that could change overall ecosystem behavior, White and Jentsch warned. Dwayne Estes presented the perfect example in the way that the predominant focus on forests in the Cumberland Plateau is at the expense of the plateau's natural grasslands and will throw the plateau's ecology as a whole out of whack. The underground seed bank can only hold on for so long in smaller and smaller sites shaded out by a tree canopy. On a much smaller scale, with a species population as reduced in scope as *Schwalbea americana*, should we focus limited conservation resources on cultivating specimens *ex situ* for outplanting, or is it better to protect and improve habitat where the species is still holding on? The latter option is preferred among those experts I met, though working both in the greenhouse and the native habitat is the best remedy, they said.

Another core principle in conservation ethics, as defined by White and Jentsch, is that we must realize that the systems are dynamic and ever-changing. The importance of disturbance by fire, floods, and/or animal grazing suggests that "ecosystem dynamics must be accepted, not resisted," these experts said. Or, to put it another way, trying to reduce the risk of fire in the short term will likely cause more intense fires over the long term. (Think of California in 2019 and 2020.) White and Jentsch cautioned: "In general, care should be taken in placing human life and property in disturbance prone ecosystem contexts."

Ultimately, the protection of our hot spots of biodiversity, of which there are so many in the South, may require the removal of river impoundments and the allowance of natural fires. White and Jentsch also remind us that some species can only be self-sustaining on a large scale—think of lions in the Serengeti or river cane ecosystems and the species they once supported in the South. "There are limits to the resilience of ecosystems," they wrote. "This is all the more challenging since decisions must be made in the absence of complete knowledge."

If we approach planet Earth or even North America as a mosaic of biodiverse patches, not unlike Annabel Renwick's small prairie plots and Dwayne Estes's larger plots, we might have the beginnings of a dynamic strategy for sustainability.

However, as the Tennessee horticulturist Carol Reese commented, "The human population is the real problem. Too many large mammals per square mile cannot be sustainable. Nature will restore balance at some point just as it does with any large mammal that abuses resources. Restoring native plants is just addressing a symptom, not the real problem."

White and Jentsch concluded with this suggestion: "The task of each generation is more like a relay race than an individual race. Each generation must pass the baton to the next; in passing on the highest possible remnant of the original diversity, they also pass along the greatest number of future options." How we preserve our precious biodiversity, unique to the wild South, is a complex equation, but the individuals we have met across six states, each working in their patches, are heroes, and they are training the next generation to make the hard choices ahead.

Epilogue *Tending the Garden*

hen I started writing this book, I was focused on climate change. I wanted to learn from people on the front lines of species conservation about the fate of our natural world in the face of the dire predictions about cascading extinctions of species ahead of us. I wanted to learn what steps people could take to help avert disaster. Fortunately, I had finished 90 percent of my research and had met with nearly all of my interview subjects before COVID-19 emerged in the spring of 2020 and the worldwide pandemic took hold.

As I write now, infection rates are spiking for the third time, and the virus is spreading without regard to state lines. The South in particular—in all its demographic, economic, and political diversity—has been hard-hit. Now I see how the pandemic is perhaps our practice round, a chance to feel what it is like to be connected to the rest of the world by a singular threat, a common fear, and a need to protect others by changing our personal habits.

Even during the period of isolation, we have experienced tremendous mutuality in facing the risk of illness and death. Just as climate change will force people in some locations (along the coasts or in the path of droughts and hurricanes) to suffer more than others, coronavirus is a wake-up call for everyone on the planet to consider what is ahead as Earth continues to warm, polar ice caps diminish, the polar vortex drops lower, and sea levels rise.

My guess is that most citizens in the Southeast, irrespective of political affiliations, would say that our preparation for the pandemic was poor. But as air traffic and rush hour commuting fell off precipitously, we have had time to reflect on our pace and practices. It's not clear that this temporary reduction in carbon emissions had much of an effect on climate change, but on a personal scale, it felt good to go six weeks on one tank of gas. Eating at home saved money and time. I cleaned out closets and downsized, completing little projects long left undone. The convenience of virtual business meetings has, I suspect, forever changed how we will work and where. Many families and individuals have gotten back out in nature, since the outdoors carries less risk of contagion.

The director of the Southeastern Grasslands Initiative, Dwayne Estes, told me he's been birdwatching and learning about different avian species in Tennessee in a way that he never took time for in the past. "I see people who have been reconnecting with nature and remembering what they had forgotten in their childhoods," he said.

On the downside, Beth Stewart of Alabama's Cahaba River Society (CRS) emailed to report that since the pandemic, some people have moved forward with destructive activities around the river. She's received complaints of unauthorized clearing and grading close to the Cahaba and on a local lake. Beth wondered if the Environmental Protection Agency's announcement that it was suspending enforcement of environmental regulations gave some people a pass to barge ahead and ask for forgiveness, not permission. "This is damage that cannot be undone," she said, "and cannot be fully repaired or mitigated."

At the same time, the virus lockdown created space for CRS to conceive an entirely new approach to environmental education, with virtual field trips, short videos, and other forms of online learning for all ages. The effort has allowed CRS to reach many more students beyond the Cahaba watershed. Early on, La'Tanya Scott, the environmental science educator for CRS, coached sixty Girl Scouts from as far away as California on how to become citizen scientists and get involved in projects to protect their local waterways. Beth said that La'Tanya presented alongside university professors and a National Aeronautics and Space Administration scientist. The girls were fascinated and kept La'Tanya talking for an hour. "The pandemic is causing us to think in new ways," Beth added, "and hopefully rebuild systems in ways that will be more equitable."

As organizations and individual households retool and revise old habits of communication, travel, recreation, and socializing, can we also protect the environment and ourselves more conscientiously? Now is the time to reconsider our values and to tend the garden differently.

Across the six states of the South represented in these pages are remarkable people who are working to preserve the diversity of plant life that distinguishes the southeastern United States. "Why would you want to lose what is particular about your own backyard?" Emily Coffey asked as we sat in her office at Atlanta Botanical Garden. "My mother and grandmother gave me the gift of seeing the world I grew up in. They taught me to notice the little differences among flowers and trees. Here, even in midtown Atlanta, we have beautiful, old-growth trees that Atlantans love and identify with," Emily said. "Plants tend to ground people and define a place. When we lose these iconic plants and trees, the sense of identity loss is huge."

Emily admits that scientists aren't trained to express emotions in their work, but increasingly she has recognized that the field must be able to appeal to citizens on a personal level to help them better understand the importance of plants to human survival. She wants to help botanists and conservationists tell their stories more powerfully.

Speaking at the Chautauqua Institution in upstate New York in the summer of 2019, Elizabeth Rush, the Pulitzer Prize finalist and author of *Rise: Dispatches from the New American Shore*, told her audience that the constant barrage of sensational news and apocalyptic headlines about the environment tend to dull public awareness and foster doubts. Up to this point, she said, "climate change is slow and place-based. You have to live somewhere for a long time to see it." But now the pace is quickening.

Jennifer Ceska, the founding coordinator of the Georgia Plant Conservation Alliance, told me that she sees a lot of fatigue right now among her conservation colleagues in the field. To sustain and support her own energy, she has done psychological strength training, learning techniques from mindfulness practices and cognitive behavioral therapy, which she has in turn shared with her colleagues.

"This is daunting work, and we keep getting kicked in the gut," she says. "We have to constantly defend why we do this work to maintain our funding, even on a college campus!" The conservation timeline is decades long, Jennifer added, "and a lot of amazing people have left the field because they couldn't sustain themselves or maintain their livelihoods."

Still, Jennifer has cause for hope. She says all her elders in Texas, where she was raised, were "fisherfolk, farmers, or garden club ladies." As a GenXer, Jennifer said the gardening heritage did not convey at first. She was a nationally competitive synchronized swimmer in high school and college and planned to go into marine biology. "There was a huge disconnect for me until I reached young adulthood and discovered my love for plants," she said.

Today she is especially encouraged by the generation behind hers. "There's a tribe of young folks coming up who know plants and use herbs medicinally. We have students at the State Botanical Garden native plant sale who come in only wanting to buy plants that support bees. I'm blown away by them. This is not a mansion-and-malls generation—they are flipping away from that. The problem is, it's hard to get the training you need to work in botany these days."

Jennifer's longtime Alliance colleague Mincy Moffett agrees. "We are experiencing a decline worldwide in the teaching of ecologically focused plant sciences at a time when we need new people more than ever. Field-based botany is seen as stodgy or so last century," he says, "but people don't understand how exciting plant science can be." Mincy and others have noted how, in academia, the largest grants tend to go toward bioengineering to support the development of food and medicines. "I appreciate genetics," Mincy said, "but there are exciting things happening in plant conservation and horticulture, too! We are dealing with the Sixth Great Extinction, and it's going to hit the fan, and we are going to be struggling with what it's going to take to keep adapting, adjusting, and surviving."

The jobs that will be possible in the field of botany in the future are something to consider. We met inspiring young people who are overcoming the obstacles to serve.

Ashlynn Smith, in Florida, was featured in *The Marjorie*, an online media outlet named for the environmentalist Marjorie Stoneman Douglas. Ashlynn has been celebrated for her dedicated "dirty work" at Deer Lake State Park, where she had been conducting a wetland habitat restoration project in the Panhandle without a drop of herbicide. Ashlynn, whom we met on the Torreya planting expedition, is supported by ABG and funded by the National Fish and Wildlife Foundation at Deer Lake.

Alabama native Noah Yawn was working in the summer between his junior and senior years at Auburn to help inventory grasslands for Dwayne Estes and SGI.

In Birmingham, La'Tanya Scott of the CRS was teaching urban children the value of a clean river as the source of water and wildlife.

Roo George-Warren was learning and teaching the conservation wisdom of the Native Americans who first managed his homeland in South Carolina.

Botanist April Punsalan of Charleston has been writing about medicinal plants and has launched a mail-order business offering Native herbal teas.

Morgan Bettcher, at the Georgia Department of Natural Resources, explained that no matter how hopeless and depressing the lectures on environmental degradation were in college and graduate school, he and his classmates somehow became more committed to engage the challenges and enter the field.

Anne Frances is the lead botanist for NatureServe, a nonprofit organization that consolidates information on all the rare species and ecosystems collected by the individual state natural heritage programs in the United States and Canada. It then makes this information available to all kinds of users—municipalities, states, nongovernmental organizations, developers, private landowners, and others—to help them make decisions about how to manage their land.

"It is the beauty of our network that we have this information available from state and national sources," Anne says. Yet as a scientist committed to the project, she has to raise her own salary and the annual funds to support two other staff members. Anne strongly believes that the Endangered Species Act is good policy but that it is woefully underfunded and often misunderstood. "There are lots of at-risk plants that are not federally listed," she added.

"The optimum situation would be for the US Fish and Wildlife Service to be able to prevent species from ever making the endangered list," she says. Resources to implement the recovery plans that are developed by FWS staff are rarely adequate. I noted that the federally endangered species discussed in this book were listed more than twenty years ago. "Yes," Anne said, "and conservationists rarely get credit for the plants that have been kept from extinction." It's hard to prove a negative.

Mincy Moffett, the state natural heritage botanist in Georgia, pointed out that while NatureServe "is the mother ship of all state natural heritage programs," it does not fund the individual programs. Support comes from the state level, and not every state is equally well funded or equipped to carry out the necessary rare species surveys and databasing. This disparity in funding levels is apparent among the state nongame programs that are responsible for the regulatory oversight and the labor-intensive, on-the-ground, recurring work of active conservation that so many species need, including prescribed burnings, which many habitats and species in the Southeast require.

We've seen in the COVID-19 crisis what a challenge it is to coordinate multiple states. We've become familiar with the layers of laws and regulatory authorities across state and local governments collaborating (or not) to control the spread of the disease. The same applies to the common challenge of preserving a region's unique biodiversity.

Yet the extensive network of state plant conservation alliances, using data from NatureServe, can and do bring together federal and state officials, utility companies, nonprofits, corporations, private gardens, and universities. These alliances allow conservationists—public and private—to leverage collected resources to survey and conserve species. "Some states, however, have trouble keeping both their natural heritage programs and nongame conservation programs funded," Mincy said, "but the better-funded state government agencies, like ours in Georgia, will have a variety of specialists on staff to manage and protect wildlife and plants."

Mincy continued: "From a general perspective, one of the great things about this country is that we have a Bill of Rights. At its founding, America embodied a new way of sharing power among individuals, but no one conceived of a time when individual rights and liberties would be emphasized at the expense of the common good as they are today."

Mincy suggested that we need a corresponding Bill of Responsibilities: "There is an expectation that if you buy land and pay taxes on it that you have a legal right to profit from it—and that's fair. But if that purchase and ownership also came with a legal responsibility to consider its importance to all humanity and living organisms on the planet, perhaps we could have a new appreciation for environmental limits on the use of private property."

Mincy knows, of course, how difficult this kind of paradigm shift would be. Private property is a sacrosanct element of American life. "For now," he said, "we have the tool of conservation easements that retire the development rights on privately owned land. Some easements are purchased, and some are donated for the greater good."

"I believe that even Endangered Species Act opponents could support taking species off the list," Mincy said, "but recovering endangered species takes money and political will. It often involves taking land out of market circulation and putting it in conservation. Less than 10 percent of the land in the eastern United States is publicly owned. This might require that the federal government subsidize entities that lose potential income from economic development to save endangered species."

The Tennessee horticulturist Carol Reese agreed: "We need to create ways that make it profitable to save species and reward ways of blending responsibility with profitability."

If the Endangered Species Act were funded at the proper level, environmentalists have argued, many species could be taken off the list. However, particular animals and plants have also been ridiculed and questioned—notably the snail darter and the spotted owl. Corporations, industries, and individuals have worked to keep species from making the list because of the complex rules and added expense involved when rare species are in the way of their development plans. Congressional lobbyists have managed to water down the act and amend the statute to serve these constituents' interests.

The kind of shift in values that Mincy Moffett suggests is profound. For Anne Frances at NatureServe, keeping up her own well-being and managing the broad scope of her conservation work on a national level requires an incremental approach, given the magnitude of the challenges. Still, she is highly motivated by her passionate colleagues. "I have never worked with a smarter, more dedicated group of people in my life," Anne said. "You don't want to let your team down."

Focusing on maintaining critical habitat, keeping rare plants healthy, and keeping them from the tipping point is the present order of business. As climate change begins to impact the globe more dramatically, the situation will become more controversial and intense.

"Resource shortages, overpopulation, and climate change may throw us into a kind of chaos that traditional projections of geopolitical power can't really address," Mincy said. "Game-playing politics will seem much less important than addressing what will be required to stay alive on the planet. How do we feed, clothe, and shelter people, and avoid mass migration?" he asked. "Protecting the planet needs to be funded as a national and global priority."

Anne Frances is a first-generation American, and she remembers moving into her family's house in Miami when she was eight years old. Her parents were Greek immigrants who settled in Florida. Soon Anne's grandparents came to visit, bearing the gift of lemon trees, an emblem of the Greek culture her family had left behind.

"That's when I learned that a sense of place is both cultural and plant-related," she said. However, it wasn't until she took a horticulture class in college that Anne learned she could become a botanist, though she had been fascinated by plants long before, as a Girl Scout. I asked Anne and the other experts I interviewed for ways in which people can help the cause of endangered species, particularly plants. Here is a list that blends their ideas and focuses on the local level, which is critical to success:

- Donate to organizations that protect biodiversity—the full suite, not just those that protect animal species.
- Get to know the conservation organizations in your community and contribute time and money to a local land trust, a botanical garden, or a native plant society.
- Join a weed warrior program and spend a day helping to remove invasive plants in wild areas, or consider helping with rare plant surveys conducted by citizen science programs.
- Reduce the size of your lawn and instead plant species in your yard

that are native to your location. Native plants will help other native creatures in your area, especially butterflies and birds, to create a better-balanced ecosystem.

- Support organic farmers in your food purchases and avoid pesticides and herbicides as much as possible. They create a ripple effect that is damaging to pollinators and possibly to your family.
- Help raise awareness of the value of plants that are native to your area with town, county, state, and federal elected officials and support the local nurseries that sell native ecotypes.
- Contribute to internships and scholarships for underrepresented populations in the conservation and environmental field, especially people of color and women.
- Take your children and grandchildren to visit your nearest botanical garden, interpretive nature museum, or state park to witness their natural heritage, hear about native and other noninvasive plants, and learn how to identify them.

Stefan Bloodworth is the former curator of the Blomquist Native Plant Garden, at Duke University. Some years ago, on a program for North Carolina Public Television, he explained the pragmatic value of native plants. "If every city found a way to integrate their natural heritage into their branding, I think it would really help municipalities and local governments take pride in the natural heritage that they've been given and use that wealth as a way to sell their particular area."

Nature becomes an ongoing part of our identity in the South when we explain to children how historically we built our houses out of pine, our furniture out of oak, and our baskets out of river cane. These human adaptations are where nature and culture meet. "To get to know, respect, and appreciate our native plants is more important now than it ever has been," Stefan added, "and children should be given a chance to know their natural history along with their cultural history."

As the premier wildflower advocate in the nation, Lady Bird Johnson once told Peter White, then director of the North Carolina Botanical Garden, why she so strongly believed in native plant gardens. "We'd like North Carolina to keep on looking like North Carolina and for Texas to go on looking like Texas," she said in her characteristic forthrightness. As this journey across the South has demonstrated, each place offers many shades of difference. Our natural assets help to distinguish and clarify who we are as peoples who were born to different kinds of soils, climates, diets, and landscapes.

The former coal-mining town of West Blocton, Alabama, managed to create a new civic identity through the annual celebration of the rare Cahaba lily without any federal endangered species designation for their treasured plant. In May 2020, however, because of COVID-19, it was not possible to host the hundreds of guests who would normally roll into town to share a potluck meal, crowd into the Lily Center on Main Street to hear Larry Davenport give his thirty-first lecture on the species, and then shuttle down to the river to see the lilies in bloom.

But West Blocton was not deterred. Lisa Buck wrote to me, "Chuck Allen is the Festival co-coordinator and has an extensive computer background. Due to his expertise and vision, our virtual Cahaba Lily Day seemed to be a huge success."

I watched the live broadcast online from my hemlock log cabin in the Blue Ridge Mountains. A group of some ten socially distanced citizens gathered in the old department store in downtown West Blocton to crown the Lily Queen (a local high school student) and present her with a scholarship. Larry Davenport once more took the stage to give a talk and show slides. Later in the afternoon a small drone with a camera flew back and forth over Hargrove Shoals on the Cahaba River to document and share the day's profusion of blooms rising from the whitewater and rocky shoals. The event gave "live streaming" a dual meaning. Lisa Buck said she missed all the good food, but the town's reputation for beauty and an innovative spirit goes on.

Acknowledgments

Unfortunately, there are countless ways to tell the story of how humanity has endangered biodiversity in the southeastern United States. Readers will inevitably ask how I selected the dozen species of plants to highlight. With the encouragement of UNC Press executive editor Elaine Maisner, who first suggested I consider writing a book on southern plants, I began talking to professionals in the field. They worked through a list of ideas with me, making suggestions and referrals. In the end, I chose species that live in diverse ecosystems, tell us something about our cultural heritage, and have brilliant and authentic champions working to protect them.

I am grateful for the early input of Rebekah Reid and Gary Peeples of the Asheville Ecological Services Field Office of the US Fish and Wildlife Service; Damon Waitt and Joanna Massey Lelekacs of the North Carolina Botanical Garden; author and public television host Tom Earnhardt; Betsy Bennett, former director of the NC Museum of Natural Sciences; Rebecca Vidra, associate dean for learning and innovation at Duke University's Nicholas School of the Environment; and longtime conservation activist Jeff Michael, North Carolina deputy secretary for natural resources.

The day I finally sat down to interview North Carolina Herbarium director Alan Weakley with my near-final list of plants, however, I had no idea what a botanical rock star he is. I would only come to recognize his colossal contributions to the field as my research in state after state revealed his influence. Alan was also a benevolent reader of the full manuscript once completed, as was Tennessee horticulturist Carol Reece, a feisty proponent of a balanced approach to the presence of native and non-native plantings in southern gardens. Despite their vast technical expertise, Alan and Carol understood my best hopes for this book: namely, to make the science accessible while avoiding gross oversimplifications. Any shortcomings on that account are totally mine.

The botanists, government officials, conservation researchers, native plant experts, and nonprofit plant advocates and volunteers across the six states I covered are already named in the narrative. They went the extra mile to share their personal stories and deserve the highest praise and thanks from all of us who live in the region. Their hard-won expertise, backbreaking labors, and daily vigilance in the field are keeping our region's green treasures from extinction. I hope that by meeting these people and plants, readers will be curious enough to learn more. There is no substitute for an in-person visit to the wildlife refuges, botanical gardens, state and national forests and parks, arboretums, and other public preserves that are represented in these pages.

Many other individuals helped me behind the scenes. Special thanks to Herbarium Collections and Outreach Administrator Charles Zimmerman and archivist Stephen Sinon at the New York Botanical Garden; Matt Candeias, host of the podcast *In Defense of Plants*; Cherokee storyteller Davy Arch; Katie Shaddix of the Cahaba River Society; Adam Griffith, director of the Revitalization of Traditional Cherokee Artisan Resources; Margaret Tyson of the Wolf Creek Trout Lily Preserve; Alabama author and photographer Marian Moore Lewis; Lee Carol Giduz, director of the Blowing Rock Art and History Museum; NC State University forestry professor Christopher Moorman; Jacqueline Bridger and Stacey Huskins of the Endangered Species Branch of the US Army, Fort Bragg; and Darrell Horton at the State Archives of Florida. I am also grateful to Jim Clark and Jason Tomberlin for a 2019 research grant from the North Caroliniana Society's Archie K. Davis Fellowship Program.

On the road, photographer Donna Campbell and I enjoyed the accommodations, sustenance, and assistance of friends old and new. Thanks to Lisa Buck, Susan Campbell, Virginia Crank, Joyce Fleming, Teri and Jimmy Milhous, Carol Misner and Ann Huckstep, Ellie Perzel, April Punsalan, Gena Rawdon, and Greg Screws.

This project would have been impossible and not nearly as much fun if Donna had not been riding shotgun and recording the details of our botanizing with her cameras. Donna's work speaks for itself, and the range of challenges in this project—inclement weather, snakes, extremes in lighting and temperature, swampy and highland hikes, wind, and other precarious natural elements—could never be adequately compensated. Thank you, Donna, for always being ready for another adventure and for helping me to parse all that we've seen together.

When seasonal logistics prohibited our ability to take photos of some of the species in bloom, the region's botanical community—tight-knit and generous—came to the rescue. Thanks to the other photographers credited in the book, who quickly gave permission for the use of their images. Special thanks to Beth Maynor Finch and Mincy Moffett, who paved the way for these additions with their email introductions.

I know my good fortune in having befriended the noted Alabama photographer Elmore DeMott, who convinced her husband Miles and friend Jody Thrasher to venture down Hatchett Creek in a canoe so that she could capture the cover photo for this book. Elmore has dedicated her artistic career to encouraging people to seek beauty through nature. For his part, Jody Thrasher plans to donate a Conservation Easement of this section of the creek to the Alabama Freshwater Land Trust in honor of his father, Dr. David Thrasher.

As always, my thanks go to stalwart reader and editor Debbie McGill, who manages to smooth my sentences, improve my word choices, and fix my occasional antecedent troubles. Thank you, DM, especially in a year when your dance card was already too full. Thanks also to Robert Anthony, Matthew Booker, Lawrence Earley, Wayne Goodall, Jill McCorkle, Jesse Pope, Marsha Warren, and Emily Wilson for support, ideas, and encouragement along the way.

At UNC Press, my gratitude extends across a dedicated staff who, during the pandemic, have had to work in isolation from each other to produce this book and others. So many hands have made it happen: executive editor Elaine Maisner, assistant managing editor Jay Mazzocchi, associate editor Andrew Winters, and acquisitions assistant Andreina Fernandez. In the production phase, I benefited from the meticulous work of senior project editor Erin Granville, copyeditor Christi Stanforth, and designer Richard Hendel. The marketing, development, and publicity teams at UNC Press have been innovative and industrious partners—especially Dino Battista, Joanna Ruth Marsland, and Allie Shay. Editorial director Mark Simpson-Vos and associate editor Cate Hodorowicz have provided essential support and technical expertise throughout. I am also grateful to the Blythe Family Fund of the University of North Carolina Press for publication assistance. Native bouquets to you all.

Bibliography

Introduction

Allison, J. R., and T. E. Stevens. "Vascular Flora of the Ketona Dolomite Outcrops in Bibb County, Alabama." *Castanea* 66, nos. 1 and 2 (2001): 154–205.

Borenstein, Seth. "Florence Could Dump Enough Rain to Fill the Chesapeake Bay." AP News, 14 September 2018. https://apnews.com/b9502efb4d2b4e749f 1d525d06661367/Florence-could-dump-enough-rain-to-fill-the-Chesapeake -Bay.

Daniel, Pete. *Toxic Drift: Pesticides and Health in the Post–World War II South.* Baton Rouge: Louisiana State University Press, 2007.

Davenport, Lawrence J., Chris Oberholster, and Brian R. Keener. "Endangered and Threatened Plants of Alabama." *Encyclopedia of Alabama*, 2 February 2018. http://www.encyclopediaofalabama.org/article/h-3241.

Finch, Bill. "The True Story of Kudzu, the Vine That Never Truly Ate the South." *Smithsonian Magazine*, September 2015. https://www.smithsonianmag.com /science-nature/true-story-kudzu-vine-ate-south-180956325/.

Lewis, Herbert J. "Little Cahaba Iron Works." *Encyclopedia of Alabama*, 14 June 2012. http://www.encyclopediaofalabama.org/article/h-1139.

Louv, Richard. *Last Child in the Woods: Saving Our Children from Nature Deficit Disorder.* Chapel Hill: Algonquin Books, 2008.

Marris, Emma. *Rambunctious Garden: Saving Nature in a Post-wild World.* New York: Bloomsbury USA, 2011. See pp. 12, 15.

McKibben, Bill. *Falter: Has the Human Game Begun to Play Itself Out?* New York: Henry Holt and Company, 2019. See p. 50.

Nash, Roderick Frazier. *Wilderness and the American Mind.* New Haven, CT: Yale University Press, 2001. See p. 1.

National Park Service. "National Park System Timeline (Annotated)." History E-Library, accessed 11 February 2021. https://www.nps.gov/parkhistory /hisnps/NPSHistory/timeline_annotated.htm.

Nowosad, Jakub, Tomasz F. Stepinski, and Pawel Netzel. "Global Assessment and Mapping of Changes in Mesoscale Landscapes: 1992–2015." *International Journal of Applied Earth Observation and Geoinformation* (June 2019).

Wilson, Edward O. *Half Earth: Our Planet's Fight for Life.* New York: W. W. Norton, 2016. See pp. 21–22, 137.

Chapter 1. Yadkin River Goldenrod and Heller's Blazing Star

Barnhart, John Henley. "The Passing of Dr. Small." *Journal of the New York Botanical Garden* 39, no. 460 (April 1938). http://www.herbarium.unc.edu /Collectors/Small.pdf.

"Biography of John Kunkel Small." Records of John Kunkel Small, Archives and Manuscripts, Mertz Library, NY Botanical Garden, accessed 11 February 2021. https://www.nybg.org/library/finding_guide/archv/small_rg4b.html.

Brendel, Frederick. "Historical Sketch of the Science of Botany in North America from 1635 to 1840." *American Naturalist* 13 (December 1879): 771.

Clary, Renee, and James Wandersee. "Our Human-Plant Connection." *Science Scope* 34, no. 8 (April–May 2011).

Cockman, Crystal. "Rediscovery of the Unique Yadkin River Wildflower." Three Rivers Land Trust, 26 August 2015. https://threeriverslandtrust.org/2015 /rediscovery-of-the-unique-yadkin-river-wildflower/.

"Dr. A. A. Heller, Veteran Botanist, Chico Techer, Dies in Vacaville." *Chico Daily Record*, 20 May 1944, 1.

Dupree, A. Hunter. *Asa Gray: American Botanist, Friend of Darwin*. Baltimore: Johns Hopkins University Press, 1988.

Goldsmith, Thomas. "Badin a Mill Village with French Flair." *Charlotte Observer*, 12 July 2011. https://www.charlotteobserver.com/living/travel/article9063623 .html.

"Green Park Inn." Historic Hotels of America, accessed 11 February 2021. https:// www.historichotels.org/hotels-resorts/green-park-inn/history.php.

"Heller's Blazing Star Factsheet." US Fish and Wildlife Service, December 2011. https://www.fws.gov/southeast/pdf/fact-sheet/hellers-blazing-star.pdf.

Morris, Edmund. *Edison*. New York: Random House, 2019. See p. 64.

Nickens, T. Edward. "North Carolina's Uwharrie Mountains." *Backpacker*, 14 February 2017. https://www.backpacker.com/stories/north-carolina-s -uwharrie-mountains.

Pace, Matthew. Interview, 21 February 2019, NY Botanical Garden, Bronx.

Rogers, David. "Blowing Rock History: Yadkin River Headwaters Brought to the Fore by Highway Construction." *Blowing Rock News*, 18 June 2015. https:// blowingrocknews.com/blowing-rock-history-yadkin-river-headwaters -brought-to-the-fore-by-highway-construction.

Small, John K., and A. A. Heller. "Flora of Western North Carolina and Contiguous Territory." *Memoirs of the Torrey Botanical Club* 3, no. 1 (1892): 1–39. www.jstor.org/stable/43391908.

Troyer, James R. "Pre-Twentieth-Century Contributors to the Botany of North Carolina." *Journal of the North Carolina Academy of Science* 119, no. 3 (2003): 111–19. www.jstor.org/stable/24336360.

Wandersee, J. H. "Plants or Animals—Which Do Junior High School Students

Prefer to Study?" *Journal of Research in Science Teaching* 23, no. 5 (May 1986): 415–26.

Wandersee, J. H., and Elisabeth E. Schussler. "Preventing Plant Blindness." *American Biology Teacher* 61, no. 2 (February 1999).

Weakley, Alan. Interview, 6 December 2018, NC Botanical Garden, Chapel Hill.

"Yadkin River Goldenrod Factsheet." US Fish and Wildlife Service, September 2013. https://www.fws.gov/asheville/pdfs/Yadkin-River-Goldenrod_Factsheet .pdf.

Chapter 2. Florida Torreya

Barlow, Connie. "Anachronistic Fruits and the Ghosts Who Haunt Them." *Arnoldia* 61, no. 2 (2001): 14–21. www.jstor.org/stable/4295484.

———. "Deep Time Lags: Lessons from Pleistocene Ecology." In *Gaia in Turmoil: Climate Change, Biodepletion, and Earth Ethics in an Age of Crisis*, edited by Eileen Crist and H. Bruce Rinker, 169. Cambridge, MA: MIT Press, 2009.

———. Email communication, 29 June 2020.

———. Torreya Guardians website. Accessed 11 February 2021. torreyaguardians .org.

Candeias, Matt. Interview with Dr. Emily Coffey. *In Defense of Plants* (podcast), episode 228: "Plant Conservation in Action," 1 September 2019. https://www .indefenseofplants.com/podcast/tag/carnivorous+plants.

Chapman, Dan. "Saving the Florida Torreya: One Goal, Two Schools of Thought on Preserving the Rare, Endangered Tree." US Fish and Wildlife Service, 22 April 2019. https://www.fws.gov/southeast/articles/saving-the-florida-torreya/.

Coffey, Emily. Interview, 24 February 2020, Atlanta Botanical Garden, Atlanta, GA.

Connolly, Patrick. "Mermaids' Lair Gets Revamp at Weeki Wachee." *Boston Herald*, 25 August 2019. https://www.bostonherald.com/2019/08/25/mermaids -lair-gets-revamp-at-weeki-wachee/.

Gray, Asa. *Scientific Papers of Asa Gray.* Vol. 2, *Essays; Biographical Sketches, 1841–1886.* Edited by Charles Sprague Sargent. Boston: Riverside, 1889. See pp. 189–90.

Kirby, Jack Temple. *Mockingbird Song: Ecological Landscapes of the South.* Chapel Hill: University of North Carolina Press, 2006. See p. 30.

Marinelli, Janet. "For Endangered Florida Tree, How Far to Go to Save a Species?" *Yale Environment 360*, 27 March 2018. https://e360.yale.edu/features/for -endangered-florida-tree-how-far-to-go-to-save-a-species-torreya.

Marris, E. "Moving on Assisted Migration." *Nature Climate Change* 1, nos. 112–13 (2008). https://doi.org/10.1038/climate.2008.86.

Pittman, Craig. "North Florida Is Gholson's Garden." *South Florida Sun Sentinel*, 10 November 2002. https://www.sun-sentinel.com/news/fl-xpm-2002-11-10 -0211100129-story.html.

Rivera Vargas, L. I., and V. Negrón-Ortiz. "Root and Soil-Borne Oomycetes (Heterokontophyta) and Fungi Associated with the Endangered Conifer, *Torreya taxifolia* Arn. (Taxaceae) in Georgia and Florida, USA." *Life: The Excitement of Biology* 1, no. 4 (2013). https://www.fws.gov/panamacity /resources/RiveraVargasand%20Negron-Ortiz.pdf.

Salustri, Cathy. "What Hurricane Michael Did to the Torreya Tree." *Tampa Bay Times*, 3 October 2019. https://www.tampabay.com/news/2019/10/03/what -hurricane-michael-did-to-the-torreya-tree/.

Schwartz, M. W. "Conservationists Should Not Move Torreya Taxifolia." *Wild Earth Forum* 10 (January 2005).

Small, John K. *From Eden to Sahara: Florida's Tragedy.* 1929. Repr., Sanford, FL: Seminole Soil and Water Conservation District, 2004.

Whitney, Ellie, D. Bruce Means, and Anne Rudloe. *Florida's Uplands.* Sarasota: Pineapple Press, 2014.

Chapter 3. Alabama Canebrake Pitcher Plant and Green Pitcher Plant

Associated Press. "Venus Flytrap Poachers Charged, Jailed." *Wilmington Star News*, 3 December 2015. https://www.starnewsonline.com/article/NC /20151203/NEWS/151209738/WM.

Bartram, William. *Travels of William Bartram.* Edited by Mark Van Doren. New York: Dover Publications, 1928.

Bullock, Clint. "NC Man Accused of Poaching Venus Flytraps Faces 73 Felonies." WECT News, 19 March 2019. https://www.wect.com/2019/03/18/nc-man -accused-poaching-venus-flytraps-faces-felonies/.

Byrd, Chuck. Interview, 6 May 2019, Bibb County, AL.

Candeias, Matt. Interview with Dr. Aaron Ellison. *In Defense of Plants* (podcast), episode 164: "Carnivorous Plants: Their Physiology, Ecology, and Evolution," 10 June 2018. http://www.indefenseofplants.com/podcast/2018/6/10/ep-164 -carnivorous-plants-their-physiology-ecology-and-evolution.

———. Interview with Dr. Jess Stephens. *In Defense of Plants* (podcast), episode 210: "Pitcher Plants: A World unto Themselves," 28 April 2019. http://www .indefenseofplants.com/podcast/2019/4/28/ep-210-pitcher-plants-a-world -unto-themselves.

Duncan, R. Scot, and Edward O. Wilson. *Southern Wonder: Alabama's Surprising Biodiversity.* Tuscaloosa: University of Alabama Press, 2013. See pp. 186, 216–17.

Ellison, Aaron. "The 'Most Wonderful Plants in the World' Are Also Some of the Most Useful Ones." *OUPblog*, Oxford University Press, 21 February 2018. https://blog.oup.com/2018/02/carnivorous-plants-useful-nature/.

Ferebee, Johanna. "Why Isn't It Illegal When Developers Destroy Venus Flytraps, and Other Reader Questions." *Port City Daily* (Wilmington, NC), 21 March

2019. https://portcitydaily.com/local-news/2019/03/21/why-isnt-it-illegal
-when-developers-destroy-venus-flytraps-and-other-reader-questions/.

Goins, Brandon. "Biodiversity Day: Venus Flytrap Preservation." NC Department
of Natural and Cultural Resources, 21 May 2019. https://www.ncdcr.gov/blog
/2019/05/21/biodiversity-day-venus-flytrap-preservation.

Rainer, David. "Rare Pitcher Plant Survives at DeSoto Park." *Courier Journal*
(Florence, AL), 18 June 2015. https://www.courierjournal.net/online_only
/article_c6c22802-15d7-11e5-9549-7be500601be0.html.

Rice, Barry. Email communication, 29 June 2020.

Stephens, Jessica D. Email communication, 23 June 2020.

Stephens, Jessica D., and Debbie R. Folkerts. "Life History Aspects of *Exyra
semicrocea* (Pitcher Plant Moth) (Lepidoptera: Noctuidae)." *Southeastern
Naturalist* 11, no. 1 (2012): 111–26. www.jstor.org/stable/41475434.

Thompson, Patrick. Interview, 11 February 2020, Davis Arboretum, Auburn, AL.

Tucker, Abigail. "The Venus Flytrap's Lethal Allure." *Smithsonian Magazine*,
February 2010. https://www.smithsonianmag.com/science-nature/the-venus
-flytraps-lethal-allure-5092361/.

Yawn, Noah. Phone interview, 3 December 2019.

———. Interview, 11 February 2020, Auburn, AL.

———. Email communication, 2 June 2020.

Chapter 4. Miccosukee Gooseberry

Bauman, Hannah. "Florida Scientists Work to Conserve Rare Gooseberry
Species." *American Botanical Council HerbalEGram* 22, no. 5 (May 2014): 1.
http://cms.herbalgram.org/heg/volume11/05May/GooseberryConservation
Efforts.html.

Catling, Paul M. "Extinction and the Importance of History and Dependence in
Conservation." *Biodiversity* 2, no. 3 (December 2011): 2–13.

Catling, Paul M., L. Dumouchel, and V. R. Brownell. "Pollination of the
Miccosukee Gooseberry (*Ribes echinellum*)." *Castanea* 63, no. 4 (1998): 402–7.
www.jstor.org/stable/4033993.

Colville, F. V. "Grossularia Echinella, A Spiny-Fruited Gooseberry from Florida."
Journal of Agricultural Research 28, no. 1 (1924): 71–74.

Conrad, Gibby. "The Life and Death of Hardy Croom, First Owner of Goodwood
Plantation." *Tallahassee Magazine*, 25 June 2012. https://www.tallahassee
magazine.com/the-life-and-death-of-hardy-croom-first-owner-of
-goodwood-plantation/.

Cox, Jim. "Long-Time Red Hills Naturalist Recognized for Contributions."
Tallahassee Democrat, 25 November 2015. https://www.tallahassee.com
/story/life/2015/11/25/long-time-red-hills-naturalist-recognized
-contributions/76318910/.

De Villegas, Rob Diaz. "Lake Miccosukee Sinkhole Hike: Floridan Aquifer Exposed!" Local Route, WFSU Public Media, 8 February 2018. https://wfsu .org/local-routes/2018-02-08/lake-miccosukee-sinkhole-hike-floridan -aquifer-exp/.

Engstrom, Todd. Phone interview, 16 April 2020.

———. Email communication, 21 July 2020.

Engstrom, Todd, and Tom Radzio. "What's Eating the Fruit of the Miccosukee Gooseberry?" *Castanea* 79, no. 1 (March 2014): 27–31.

Fairchild, David. *The World Was My Garden: Travels of a Plant Explorer.* New York: Charles Scribner's Sons, 1938.

Fishman, Gail. *Journeys through Paradise: Pioneering Naturalists in the Southeast.* Tallahassee: University Press of Florida, 2017.

———. Interview, 13 February 2020, St. Marks Wildlife Refuge, Wakulla County, FL.

———. Email communication, 21 April 2020.

"Frederick Vernon Coville Obituary." *Science* 85, no. 2203 (19 March 1937): 280–82.

Greipsson, Sigurdur, and Antonio DiTommaso. "Invasive Non-native Plants Alter the Occurrence of Arbuscular Mycorrhizal Fungi and Benefit from This Association." *Ecological Restoration* 24, no. 4 (December 2006).

Harper, Roland M. "A Botanically Remarkable Locality in the Tallahassee Red Hills of Middle Florida." *Torreya* 25, no. 3 (May–June 1925): 45–54.

———. "Persistence of Exotic Plants under Forest Conditions." *Torreya* 31, no. 1 (1931): 1–7. www.jstor.org/stable/40596710.

"Harper's Beauty." US Botanic Garden (Washington, DC), accessed 8 February 2021. https://m.usbg.gov/plants/harpers-beauty.

Leopold, Aldo. *A Sand County Almanac with Essays on Conservation from Round River.* New York: Ballentine Books, 1970. See p. 118.

"Letchworth-Love Mounds Archeological State Park." Florida State Parks, October 2015. https://www.floridastateparks.org/parks-and-trails/letchworth -love-mounds-archaeological-state-park/history.

McGlynn, Sean. *Leon County Lakes 2006.* McGlynn Laboratories Inc. Reports. http://www.mcglynnlabs.com/florida-lab-reports/.

McMillan, Patrick. "Stevens Creek Heritage Preserve." *Journal of the South Carolina Native Plant Society* (Winter 2002).

Miccosukee Tribe of Indians of Florida. "History." Accessed 8 February 2021. https://tribe.miccosukee.com.

Miller, James H. *Nonnative Invasive Plants of Southern Forests: A Field Guide for Identification and Control.* General Technical Report SRS-62. Asheville, NC: US Department of Agriculture, Forest Service, Southern Research Station, 2003. https://www.srs.fs.usda.gov/pubs/gtr/gtr_srs062/.

Negrón-Ortiz, Vivian. "Breeding System, Seed Germination and Recruitment of a

Threatened, Southeastern U.S. Endemic, *Ribes echinellum* (Grossulariaceae)." *Rhodora* 120, no. 982 (2018): 99–116.

———. "*Ribes echinellum* (Miccosukee gooseberry) 5-year Review: Summary and Evaluation." US Fish and Wildlife Service, Panama City, FL, 18 June 2015.

———. Phone interview, 15 November 2018.

Odenwald, Neil G., and James R. Turner. *Identification, Selection, and Use of Southern Plants: For Landscape Design.* 4th ed. Baton Rouge: Claitor's Law Books and Publishing, 2006.

Oleas, Nora H., Eric J. B. von Wettbert, and Vivian Negrón-Ortiz. "Population Genetics of the Federally Threatened Miccosukee Gooseberry (*Ribes echinellum*), an Endemic North American Species." Springer Science+ Business Media, 23 February 2014. https://www.fws.gov/panamacity /resources/Ribesgeneticsarticle.pdf.

Perkins, Kent. "Object 54: Florida Yew." University of Florida Herbarium Collection, Florida Museum of Natural History, accessed 8 February 2021. https://www.floridamuseum.ufl.edu/100years/florida-yew/.

Pritzker, Barry. *A Native American Encyclopedia: History, Culture, and Peoples.* Oxford: Oxford University Press, 2000. See pp. 389–91.

Radford, A. E., and H. E. Ahles. "Species New to the Flora of South Carolina." *Journal of the Elisha Mitchell Scientific Society* 75 (1959): 35–43.

Redfearn, D. H. "The Steamboat *Home*: Presumption as to Order of Death in a Common Calamity." *Florida Law Journal* 9, no. 5 (May 1935): 405–24. Excerpts available at http://herbarium.unc.edu/Collectors/Croom_HB.htm.

Shores, Elizabeth Findley. *On Harper's Trail: Roland McMillan Harper, Pioneering Botanist of the Southern Coastal Plain.* Athens: University of Georgia Press, 2010.

Slapcinsky, J. L., D. Gordon, and E. S. Menges. "Responses of Rare Plant Species to Fire in Florida's Pyrogenic Communities." *Natural Areas Journal* 30 (2010): 4–19.

Small, John K. "Land of the Question Mark: Report on Exploration in Florida in December 1920." *Journal of the New York Botanical Garden* 24 (1923): 1–23.

Stone, Daniel. *The Food Explorer: The True Adventures of the Globe Trotting Botanist Who Transformed What America Eats.* New York: Dutton, 2018.

"Success Story: Taxol (NSC 125973)." National Cancer Institute, accessed 22 February 2021. https://dtp.cancer.gov/timeline/flash/success_stories/S2 _Taxol.htm.

"Suncoast Connector Task Force." Florida Department of Transportation, 2020. https://floridamcores.com/suncoast-connector-task-force/.

"Suncoast Connector Toll Road Threatens Red Hills and Big Bend." Tall Timbers, accessed 8 February 2021. https://talltimbers.org/suncoast-connector/.

Troyer, James R. "Hardy Bryan Croom (1797–1837), Pioneer Botanist of North Carolina and the Southeast." *Journal of the North Carolina Academy of Science* 118, no. 2 (Summer 2002): 63–69.

US Science Service. "Remarkable New Berry Discovered in Florida." *Daily Science News Bulletin* 184A (1924): 1–2.

Whetstone, R. David. "Notes on Croomia Pauciflora (*Stemonaceae*)." *Rhodora* 86, no. 846 (1984): 131–37. www.jstor.org/stable/23314301.

Young, Susan H. "Miccosukee Gooseberry." Endangered Species Articles pages, South Carolina Wildlife Federation website, 2 April 2004. http://www.scwf .org/miccosukee-gooseberry.

Chapter 5. Shoals Spider Lily (Cahaba Lily)

Baldeck, Brett. "100,000 Gallons of Sewage Spill into Catawba River during Flooding." Fox 46 Charlotte, 12 June 2019. https://www.fox46charlotte.com /news/100000-gallons-of-sewage-spill-into-catawba-river-during-flooding.

Buck, Lisa. Interview, 9 February 2020, West Blocton, AL.

———. Email communication, 25 July 2020.

Cahaba River Society. Official website, accessed 9 February 2021. https:// cahabariversociety.org.

"Catawba River North Carolina South Carolina." American Rivers, accessed 8 February 2021. https://www.americanrivers.org/river/catawba-river/.

Davenport, Lawrence J. "The Cahaba Lily." *Alabama Heritage* 16 (Spring 1990): 24–29.

———. "Cahaba Lily." *Encyclopedia of Alabama*, 18 August 2017. http://www .encyclopedia ofalabama.org/article/h-967.

———. "The Cahaba Lily: Its Distribution and Status in Alabama." *Journal of the Alabama Academy of Sciences* 67 (October 1996): 222–33.

———. Interview, 10 February 2020, Samford University, Birmingham, AL.

Duncan, R. Scot, and Edward O. Wilson. *Southern Wonder: Alabama's Surprising Biodiversity*. Tuscaloosa: University of Alabama Press, 2014.

"Faculty Archive: William Hardy Eshbaugh." Southern Illinois University Department of Plant Biology, Carbondale, accessed 10 February 2020. https:// plantbiology.siu.edu/about/history/history-faculty/eshbaugh.php.

Golladay, Stephen W., and Kevin McIntyre. "Flora and Fauna of the Flint River." Sherpa Guides: The Natural Georgia Series, accessed 8 February 2021. https:// www.sherpaguides.com/georgia/flint_river/wildnotes/index.html.

Hamblin, James. "A Racial History of Drowning." *Atlantic*, 11 June 2013. https:// www.theatlantic.com/health/archive/2013/06/a-racial-history-of-drowning /276748/.

Haynes, David. "Rivers in Bloom: Festivals Celebrate the Lovely, Fragile Cahaba

Lily." *Alabama Living Magazine*, 30 March 2015. https://alabamaliving.coop /article/rivers-in-bloom/.

Henderson, Gary. "Bathed in Blossoms Landsford Canal State Park History Soaks Canal Park." GoUpstate.com, 18 May 2002. https://www.goupstate.com /article/NC/ 20020518/News /605168452/SJ.

Holbrooks, Nick. Interview, 25 February 2020, Elbert County, GA.

———. Email communication, 28 May 2020.

Jones, Myrtle. Interview, 9 February 2020, West Blocton, AL.

Kaetz, James P. "West Blocton." *Encyclopedia of Alabama*, 16 August 2018. http:// www.encyclopediaofalabama.org/article/h-3357.

Katawba Valley Land Trust. Official website, accessed 8 February 2021. http:// www.kvlt.org/.

"Landsford Canal State Park History and Interpretation." South Carolina State Parks, accessed 8 February 2021. https://southcarolinaparks.com/landsford -canal/history-and-interpretation#jump.

Lanier, Wayne. Interview, 30 May 2019, Landsford Canal State Park, SC.

Nijhuis, Michelle. "The Cahaba: A River of Riches." *Smithsonian Magazine*, August 2009. https://www.smithsonianmag.com/science-nature/the-cahaba -a-river-of-riches-34214889 /#1iFceT5aEm7EqAEz.99.

Overton, Dr. Anthony. Phone interview, 24 April 2020.

Scott, La'Tanya. Email communication, 18 June 2020.

Seabrook, Charles. "Rare Shoals Spider-lily Is a Stunning Sight." *Atlanta Journal-Constitution*, 30 May 2014. https://www.ajc.com/lifestyles/rare-shoals-spider -lily-stunning-sight/7dWehueDNQn1dXlBZNUNwI/.

Stewart, Beth, Randy Haddock, and Gordon Black. Interview, 10 February 2020, Cahaba River Society offices, Birmingham, AL.

Wayland, Michael. "Mercedes-Benz Reopened an Alabama Auto Plant. Here's How it Happened." CNBC, 30 April 2020. https://www.cnbc.com/2020/04/30 /mercedes-benz-reopened-an-alabama-auto-plant-heres-how-it-happened .html.

West Blocton Improvement Committee. *Coal Miner's Vittles: A Collection of Recipes*. Kearney, NE: Morris Press Cookbooks, 2012. See p. 4.

Wright, Erica. "Cahaba River Society's La'Tanya Scott: Sharing Love of Nature with Children." *Birmingham Times*, 18 October 2018. https://www .birminghamtimes.com/2018/10/cahaba-river-societys-latanya-scott-sharing -love-of-nature-with-children/.

Chapter 6. Morefield's Leather Flower

Brown, Alyssa J., Phil S. Allen, Greg V. Jolley, and Ryan J. Stewart. "The Downward Trend in Postsecondary Horticulture Program Availability between 1997 and

2017." *HortTechology* 29, no. 4 (20 June 2019). https://journals.ashs.org
/horttech/view/journals/horttech/ 29/4/article-p417.xml.

Cook, Tracy. Email communication, 13 June 2019.

———. Interview, 13 June 2019, Huntsville Botanical Garden, Huntsville, AL.

Crabtree, Todd. "Search for Populations of *Clematis morefieldii* Kral in
Tennessee." US Fish and Wildlife Service, March 2011.

———. "Vanderbilt's Robert Kral Fifth Member of TNPS [Tennessee Native Plant
Society] Hall of Fame." *Newsletter of the Tennessee Native Plant Society* 36,
no. 4 (2012): 5.

———. Email communication, 18 June 2019.

Cressler, Alan. "Photographer's Notes on *Clematis morefieldii*." Lady Bird
Johnson Wildflower Center Plant Images Collection, accessed 8 February
2021. https://www.wildflower.org/gallery/result.php?id_image=44705.

Estes, Dwayne, and Chris Fleming. "*Clematis morefieldii* (Ranunculaceae) New to
Tennessee." *SIDA, Contributions to Botany* 22, no. 1 (2006): 821–24. www.jstor
.org/stable/41968647.

———. "Is Over-Reliance of Using Herbarium Specimens for Taxonomic Studies
Leading Us to Underestimate Southeastern Plant Diversity? (Part 1 of 2)."
Natural History, Flora, and Vegetation of the Southeastern US (blog), 5 May
2015. http://southeastveg.blogspot.com/ 2015/05/is-over-reliance-of-using
-herbarium.html.

———. "Untangling the Viny Viornas: A Case of How We've Underestimated
Biodiversity in Architecturally Complex Genera (Part 2 of 2)." *Natural History,
Flora, and Vegetation of the Southeastern US* (blog), 11 November 2015. http://
southeastveg.blogspot.com.

Evans, Jon. "Domain Flora Published in Castanea." Plant Ecology and
Conservation Lab, Sewanee Herbarium, 23 September 2016. https://www
.evanslab.org/single-post/2016/09/23/Domain-Flora-Published-in-Castanea.

Evans, Jon, Angus Pritchard, and Lillian Fulgham. Interview, 13 June 2019,
Sewanee Herbarium, University of the South, Sewanee, TN.

Kral, Robert. "A New 'Viorna' Clematis from Northern Alabama." *Annals of the
Missouri Botanical Garden* 74, no. 3 (1987): 665–69. www.jstor.org/stable
/2399332.

Morefield, James D. Email communication, 14 June, 11 December 2019.

Paek, Matthew. "Morefield's Leather Flower." *Encyclopedia of Alabama*,
5 February 2019. http://www.encyclopediaofalabama.org/article/h-4073.

Pritchard, Angus. "A Summer in Sewanee with the Herbarium Fellows." *Sewanee
Plant Press* 23, no. 4 (Autumn 2019).

Steinmann, Hali. "Plant of the Day: Our Very Own Endangered Species."
Sewanee Herbarium, 12 June 2013. https://sewaneeherbarium.wordpress
.com/2013/06/12/plant-of-the-day-our-very-own-endangered-species/.

US Fish and Wildlife Service. "Endangered and Threatened Wildlife and Plants: Determination of *Clematis morefieldii* (Morefield's Leather Flower) to Be an Endangered Species." *Federal Register* 57 (1998): 21562–34420.

———. "Recovery Plan for Morefield's Leather Flower (*Clematis morefieldii*)." Atlanta, 3 May 1994.

Chapter 7. Michaux's Sumac

Bender, Steve. "Don't Fear Sumac." The Grumpy Gardener, *Southern Living*, 20 October 2010. https://www.southernliving.com/garden/grumpy-gardener /dont-fear-sumac.

Bettcher, Morgan. Interview, 25 and 26 February 2020, northeast Georgia.

Birchfield, Candis. "From Economist to Biologist, How One COSAM Alumnus Made a Transformation toward Conservation." Auburn University College of Sciences and Mathematics, 22 June 2017. https://www.auburn.edu/cosam /news/articles/ 2017/06/from-economist-to-biologist-how-one-cosam -alumnus-made-a-transformation-toward-conservation.htm.

Borenstein, Seth. "Butterfly on a Bomb Range: Endangered Species Acts at Work." AP News, 18 November 2019. https://apnews.com/5283261bea5d4a 7ea66e886410436bfc.

Ceska, Jennifer. Phone interview, 10 June 2020.

"Georgia Plant Conservation Network Earns National Award." Georgia Department of Natural Resources, Wildlife Resources Division, 14 September 2016. www.georgiawildlife.com/news.aspx.

"GPCA Safeguarding Policy Statement, 2008." Georgia Plant Conservation Alliance, accessed 10 February 2021. https://botgarden.uga.edu/wp-content /uploads/2017/01/GPCA-Safeguarding-Policy-Statement-.pdf.

Holbrooks, Nick. Interview, 25 February 2020, Elbert County, GA.

———. Email communication, 28 May 2020.

Miller, Susan. "Michaux's Sumac (*Rhus michauxii*)." US Fish and Wildlife Service, Raleigh Ecological Services Field Office, November 2020. https://www.fws .gov/raleigh/ species/es_michauxs_sumac.html.

Moffett, Mincy. Email communication, 13 November 2018, 16 December 2019, 18 June 2020.

———. Interview, 25 and 26 February 2020, northeast Georgia.

Murdock, Nora A., and Julie Moore. "Recovery Plan for Michaux's Sumac (*Rhus michauxii*)." US Fish and Wildlife Service, Atlanta, 1993.

Price, Jay. "Military Bases Serve as Safe Haven for Endangered Species." *All Things Considered*, National Public Radio, 15 September 2016. https://www .npr.org/2016/09/15/494127912/military-bases-serve-as-safe-haven-for -endangered-species.

Rembert, David H. "The Carolina Plants of Andre Michaux." *Castanea* 44, no. 2 (1979): 65–80. www.jstor.org/stable/4032704.

Shearer, Lee. "Botanists Play the Mating Game with Dwarf Sumac." *Online Athens: Athens Banner Herald*, 15 February 2010.

———. "Love Blossoms between Long-Separated Plants, Babies May Be on the Way." *Online Athens: Athens Banner Herald*, 29 September 2012. https://www.onlineathens.com/ article/20120929/NEWS/309299941.

———. "Unprecedented Award for Georgia Plant Conservation Alliance." *Online Athens: Athens Banner Herald*, 15 September 2016. http://onlineathens.com/mobile/2016-09-15/unprecedented-award-georgia-plant-conservation-alliance#.

Sullivan, Randall. "American Stonehenge: Monumental Instructions for the Post-apocalypse." *Wired*, 20 April 1989. https://www.wired.com/2009/04/ff-guidestones/.

"Super-Tough Seed Coat Keeps Michaux's Sumac on Critically Endangered List." *Smithsonian Insider*, 13 October 2011. https://insider.si.edu/2011/10/endangered-sumac/.

Wheatley, Thomas. "Flashback: Karl Wallenda's High-Wire Walk across Tallulah Gorge, 1970." *Atlanta Magazine*, 5 October 2018. https://www.atlantamagazine.com/news-culture-articles/flashback-karl-wallendas-high-wire-walk-across-tallulah-gorge-1970/.

Williams, Charlie. "Michaux's Sumac *Rhus Michauxii*." Endangered Species Articles pages, South Carolina Wildlife Federation website, 16 June 2003. http://www.scwf.org/michauxs-sumac.

Williams, Pandra. Interview, 24 February 2020, Beech Hollow Farms, Lexington, GA.
———. Email communication, 22 May 2020.

Chapter 8. River Cane

Anderson, M. K., and T. Oakes. "Plant Guide for Giant Cane *Arundinaria gigantea*." USDA-Natural Resources Conservation Service, National Plant Data Team, Greensboro, NC, 2011. https://mafiadoc.com/giant-cane-arundinaria-gigantea-plant-guide-usda-plants-_5a0a37301723dd416d8e13e4.html.

Audubon, John J. *Ornithological Biography, or an Account of the Habits of the Birds of the United States of America*. Edinburgh: Adam Black, 1833.

Cirtain, Margaret C. "Identifying Native Bamboos." *Journal of the South Carolina Native Plant Society* (Winter 2010): 8–11.

Cook, Joe. *Flint River User's Guide*. Athens: University of Georgia Press, 2017. See p. 103.

Cozzo, David. Interview, 26 April 2019, EBCI Center, Cherokee, NC.

Ellison, George. "Nature Journal: Bamboo and Hill Cane." *Asheville Citizen Times*,

13 January 2015. https://www.citizen-times.com/story/sports/outdoors/2016
/01/13/nature-journal-wncs-bamboo-often-hill-cane/78732066/.

Fariello, M. Anna. *From the Hands of Our Elders: Cherokee Traditions.*
Cheltenham, UK: History Press, 2011.

Fulcher, B. "Raising Cane in Tennessee." *Tennessee Conservationist,* July–August
1999, 11–15.

Lossie, Ramona. Interview, 20 June 2019, Qualla Arts and Crafts Mutual,
Cherokee, NC.

Hawkins, Benjamin. *The Collected Works of Benjamin Hawkins, 1796–1810.* Edited
by Thomas Foster. Tuscaloosa: University of Alabama Press, 2003. See pp. 36,
64, 126.

Henderson, Jill. "America's Native Bamboo: History and Ecology." *Show Me Oz*
(blog), 8 July 2015. https://showmeoz.wordpress.com/2015/07/08/americas
-native-bamboo-part-i-history-and-ecology/.

———. "America's Native Bamboo: Part II—Identification and Culture." *Show
Me Oz* (blog), 15 July 2015. https://showmeoz.wordpress.com/2015/07/15
/americas-native-bamboo-part-ii-identification-culture/.

Jackson, Jason Baird. "Containers of Tradition: Southeastern Indian Basketry."
Gilcrease Museum, 18 March 2016. https://collections.gilcrease.org/articles
/article-containers-tradition-southeastern-indian-basketry.

Klaus, Nathan. Telephone interview, 19 December 2019.

Klaus, Nathan, and Joyce M. Klaus. "Evaluating Tolerance of Herbicide and
Transplantation by Cane (a Native Bamboo) for Canebrake Restoration."
Restoration Ecology 19, no. 3 (September 2009): 344–50.

Kucharksi, Sarah. "Giving Art a Hand." *Smoky Mountain News,* 14 December
2005. https://www.smokymountainnews.com/archives/item/15319-giving-art
-a-hand.

Lawson, John. *A New Voyage to Carolina.* 1709. Repr., Chapel Hill: University of
North Carolina Press, 1967. See pp. xv, 107, 185, 218, 229.

Lovovikov, Maxim, Yiping Lou, Dieter Schoene, and Raya Widenoja. "The Poor
Man's Carbon Sink: Bamboo in Climate Change and Poverty Alleviation."
Rome: Forestry Department, Food and Agriculture Organization of the United
Nations, 2009.

Mississippi State University. Rivercane website, 31 July 2008. https://www
.rivercane.msstate.edu.

Moffett, Mincy. Email communication. 1 June 2020.

Norris, David A. "Canebrakes." In *Encyclopedia of North Carolina,* edited by
William Powell. Chapel Hill: University of North Carolina Press, 2006. https://
www.ncpedia.org/canebrakes.

Noss, Reed F. *Forgotten Grasslands of the South: Natural History and
Conservation.* Washington, DC: Island Press, 235.

Platt, Steven G., Christopher G. Brantley, and Thomas R. Rainwater. "Canebrake Fauna: Wildlife Diversity in a Critically Endangered Ecosystem." *Journal of the Elisha Mitchell Scientific Society* 117, no. 1 (2001): 1–19.

———. "Canebrakes: Bamboo Forests of the Southeast." *Wild Earth* (Spring 2002): 38–45.

Power, Susan. *Art of the Cherokee*. Athens: University of Georgia Press, 2007. See p. 29.

Queen, Frank G. "River Cane of the Southern Appalachians." *Smoky Mountain Living*, 1 September 2009. https://www.smliv.com/outdoors/river-cane-of -the-southern-appalachians/.

Reid, Rebekah. Email communication, 5 November 2018.

Roosevelt, Teddy. "In the Louisiana Canebrakes." *Scribners Magazine* 42 (1908): 47.

South River Watershed Alliance. Official website, accessed 8 February 2021. https://www.southriverga.org.

State of Georgia. "Original and 1895 Counties and Land Lot Districts" (maps). In personal collection of Nathan Klaus.

Stewart, Mart A. "From King Cane to King Cotton: Razing Cane in the Old South." *Environmental History* 12, no. 1 (2007): 59–79. www.jstor.org/stable/2547303.

Triplett, J. K., A. S. Weakley, and L. G. Clark. "Hill Cane (*Arundinaria appalachiana*), a New Species of Bamboo from the Southern Appalachian Mountains." *SIDA, Contributions to Botany* 22 (2006): 79–95.

Wescott, David, ed. *Primitive Technology II: Ancestral Skill*. Layton, UT: Gibbs, 2001.

Chapter 9. Schweinitz's Sunflower

Barden, Lawrence S. "Historic Prairies in the Piedmont of North and South Carolina, USA." *Natural Areas Journal* 17, no. 2 (1997): 149–52. www.jstor.org /stable/43911660.

Bates, Moni. "A Moravian Minister and His Rare Sunflower." *Landmark: A Newsletter of the LandTrust for Central North Carolina*, Spring 2001.

Bloom, Sean. Phone interview, 2 July 2020.

———. Email communication, 15 July 2020.

Brammer, John Paul. "American Two Spirit Fights to Keep Tribe's Language Alive." NBC News, 8 May 2017. https://www.nbcnews.com/feature/nbc-out /native-american-two-spirit-fights-keep-tribe-s-language-alive-n755471.

"Catawba Students Help Census the Endangered Schweinitz's Sunflower in Two N.C. Counties." College News section, Catawba College website, 6 November 2017. https://catawba.edu/news-events/news/college-news/catawba -students-help-census-endangered-schweinitzs-sunflower-two-nc-counties/.

"Coal Ash Cleanup Settlement Announced." *Observer News Enterprise*, 3 January

2020. https://www.observernewsonline.com/content/coal-ash-cleanup
-settlement-announced.

Fields, Seven E. "Schweinitz's Sunflower Helianthus Schweinitzii Torrey and Gray (Asterales: Asteraceae) in Upper Piedmont South Carolina." *Journal of the South Carolina Academy of Science* 4, no. 1 (Fall 2007): 27–32.

George-Warren, DeLesslin. Phone interview. 19 October 2019.

———. Interview, 29 November 2019, Catawba Reservation, Fort Mill, SC.

Hansen, Janice. "A Family Affair: Lewis David de Schweinitz's Drawings of Fungi." *Chapel Hill Rare Book Blog*, Rare Book Collection at UNC Chapel Hill, 28 September 2016, https://blogs.lib.unc.edu/ rbc/2016/09/28/a-family-affair -lewis-david-de-schweinitzs-drawings-of-fungi/.

"A Home for Imperiled Plants." NC Zoo, 1 August 2018. https://www.nczoo.org /conservation/zoo/home-imperiled-plants.

Hunt Institute for Botanical Documentation. "Braun, E. Lucy 1889–1971." Archives section, Hunt Institute for Botanical Documentation website, accessed 8 February 2021. https://www.huntbotanical.org/archives/detail.php?33.

Johnson, Walter Rogers. *A Memoir of the Late Lewis David Von Schweinitz, P. D. with a Sketch of His Scientific Labours, Read before the Academy of Natural Sciences of Philadelphia, May 12, 1835*. Philadelphia: William P. Gibbons, 1835.

Karakehian, J. M., W. R. Burk, and D. H. Pfister. "New Light on the Mycological Work of Lewis David von Schweinitz." *IMA Fungus* 9, no. 17 (24 June 2018). https://imafungus.biomedcentral. com/track/pdf/10.1007/BF03449476.

Lampel, Lenny. Phone interview, 15 November 2018.

Matthews, James F., Lawrence S. Barden, and Christopher R. Matthews. "Corrections of the Chromosome Number, Distribution and Misidentifications of the Federally Endangered Sunflower, *Helianthus schweinitzii* T&G." *Journal of the Torrey Botanical Society* 124, no. 2 (April–June 1997): 200.

McCormick, Carol Ann. "Paul Otto Schallert, M.D." *Collectors of the UNC Herbarium*, last updated 31 January 2017. http://herbarium.unc.edu/Collectors /Schallert_Paul.htm.

Rogers, Donald P. "L. D. De Schweinitz and Early American Mycology." *Mycologia* 69, no. 2 (1977): 223–45.

Schallert, P. O. "Schweinitz's Collecting-Ground in North Carolina." *Bartonia* 16 (1934): 8–12.

"Schweinitz's Sunflower Factsheet." US Fish and Wildlife Service, Asheville Field Office, December 2011. https://www.fws.gov/southeast/pdf/fact-sheet /schweinitzs-sunflower.pdf.

Siler, Robert. "Schweinitz's Sunflower—*Helianthus schweinitzii*." Endangered Species Articles pages, South Carolina Wildlife Federation website, 16 June 2003. http://www.scwf.org/schweinitzs-sunflower.

Smith, Kate Rauhauser. "History-Makers: Lewis de Schweinitz." *Winston-Salem Monthly*, 31 October 2017. https://journalnow.com/winstonsalemmonthly /history-makers-lewis-de-schweinitz/article_56d9acf6-b989-11e7-88b6 -2feobdf9eefo.html.

South Carolina Plant Conservation Alliance. "Scheweinitz's Sunflower Rescue." *SCPCA News*, 31 May 2018. https://scplantconservation.org/2018/05/31 /schweinitzs-sunflower-rescue/.

US Fish and Wildlife Service. "Schweinitz's Sunflower Recovery Plan." Atlanta, 1994.

———. "Schweinitz's Sunflower (*Helianthus schweinitzii*) 5-Year Review: Summary and Evaluation." Asheville, NC: Asheville Field Office, 2010.

Ververka, Amber. "At 80, Biologist Matthews Still Stalks the Threatened Landscapes." *Charlotte Urban Institute Newsletter*, 2 February 2016. https:// plancharlotte.org/story/biologist-matthews-still-stalks-threatened -landscapes.

———. "Piedmont Prairies Offer Glimpses of Region's Distant Past." *Charlotte Urban Institute Newsletter*, 24 April 2012. https://ui.uncc.edu/story/piedmont -prairies-offer-glimpses-regions-distant-past.

Chapter 10. American Chaffseed

"American Chaffseed." Drawn from *Beacham's Guide to the Endangered Species of North America* for *Encyclopedia.com*, 20 December 2019. https://www .encyclopedia.com/environment/science-magazines/american-chaffseed.

"American Chaffseed / *Schwalbea americana* Factsheet." US Fish and Wildlife Service, accessed 10 February 2021. https://www.fws.gov/southeast/wildlife /plants/american-chaffseed/.

Bloodworth, Stefan. "Evolution of a Conservation Strategy." *Flora* (Friends of Sarah P. Duke Gardens Magazine), no. 61 (2017): 8–11.

Cannon, Brandi. "The Conservation of Endangered American Chaffseed *Schwalbea americana*: Video from the 2017 Master's Synthesis Competition." Presentation given at Columbia University, 4 May 2017, YouTube video, 5.46 minutes. https://www.youtube.com/watch?v=FEIA3kNAyhM.

Cockrell, Joe, and April Punsalan, "Recovery Progress for the American Chaffseed." US Fish and Wildlife Service, South Carolina Ecological Services Field Office, 27 June 2017. https://www.fws.gov/southeast/articles/recovery -progress-for-the-american-chaffseed/.

"Coffey, Linderman Honored for Conservation." *Crossville Chronicle*, 15 June 2017.

Estes, Dwayne. "The Southeastern Grasslands Initiative." Lecture at Clinton School of Public Service, University of Arkansas, 5 November 2018, YouTube

video, 46.59 mins. https://www.youtube.com/watch?v=ncmLFvpSMCU
&feature=youtu.be.

———. "The Southeastern Grasslands Initiative: Bringing Chicago-Style
Conservation to the South." Southeastern Grasslands Initiative, 1 November
2018. https://www.segrasslands.org/blog/2018/6/18/chicago-style-full-story.

———. Phone interview, 19 June 2020.

———. Interview, 24 June 2020, Coosa Wildlife Management Area, Crossville, TN.

Glitzenstein, J., Danny Gustafson, Johnny Stowe, Donna Streng, D. Bridgman,
Jennifer Fill, and Jason Ayers. "Starting a New Population of *Schwalbea
americana* on a Longleaf Pine Restoration Site in South Carolina." *Castanea*
81 (2016): 302–13.

Grimm, William C. *Indian Harvests*. New York: McGraw-Hill, 1974.

Marinelli, Janet. "Forgotten Landscapes: Bringing Back the Rich Grasslands of
the Southeast." *Yale Environment 360*, 20 June 2019. https://e360.yale.edu
/features/forgotten-landscapesbringing-back-the-rich-grasslands-of-the
-southeast.

McCormick, Carol Ann. "What Goes around Comes Around." *NCBG Newsletter*,
November–December 2009, 10.

NatureServe. "Schwalbea americana." NatureServe Explorer, accessed 10
February 2021. https://explorer.natureserve.org/Taxon/ELEMENT_GLOBAL
.2.144235/Schwalbea_americana.

Peters, Dana. "American Chaffseed (*Schwalbea americana*) Recovery Plan." US
Fish and Wildlife Service, Region Five, Hadley, MA, 29 September 1995.

Punsalan, April (moderator). American chaffseed conference call, 6 February
2020.

Pyne, Stephen. *Fire in America: A Cultural History of Wildland and Rural Fire*.
Seattle: University of Washington Press, 1997.

Reese, Carol. Email communication, 17 November 2020.

Renwick, Annabel. Phone interview, 16 July 2020.

———. Email communication, 20 July 2020.

Ross, James. *The Life and Times of Elder Reuben Ross*. Philadelphia: Grant,
Faires & Rodgers, 1882. See p. 215.

Semple, J. C. "*Symphyotrichum estesii* (Asteraceae: Astereae), a New Species of
Viruloid Aster from Tennessee." *Phytoneuron* 36 (16 October 2019).

"Shortleaf Pine Restoration Plan Executive Summary." Shortleaf Pine Initiative:
Restoring an American Forest Legacy, accessed 10 February 2021. http://
shortleafpine.net/shortleaf-pine-initiative/shortleaf-pine-restoration
-plan.

Thwaites, Reuben Gold, ed. *Early Western Travels, 1748–1846, Journal of Andre
Michaux*. Cleveland: A. H. Clark Company, 1904. See pp. 45–46.

TVA Newsroom. "Saving America's Great Southern Grasslands." The Chattanoogan.com, 10 September 2019. https://www.chattanoogan .com/2019/9/10/396000/Saving-Americas-Great-Southern-Grasslands.aspx.

"What Is Lidar?" National Ocean Service, National Oceanic and Atmospheric Administration, accessed 10 February 2021. https://oceanservice.noaa.gov /facts/lidar.html.

White, Peter S., and Anke Jentsch. "Developing Multipatch Environmental Ethics: The Paradigm of Flux and the Challenge of a Patch Dynamic World." *Silva Carelica* 49 (1 January 2005).

Whitehead, Van. "American Chaffseed—*Schwalbea americana*." Endangered Species Articles pages, South Carolina Wildlife Federation website, 3 April 2003. http://www.scwf.org/american-chaffseed.

Wright, Amy. "The Preacher of the Prairie." *Minding Nature* 12, no. 3 (Fall 2019): https://humansandnature.org/fall-2019.

Epilogue: Tending the Garden

Brown, Hanna. "Florida's Environment: 8 #FloridaWomen Who Work 'Dirty Jobs' in Defense of Our State." *The Marjorie* (blog), 7 January 2018. https:// themarjorie.org/2018/01/07/ floridas-environment-8-women-with-dirty-jobs/.

Buck, Lisa. Email communication, 12 May 2020.

Ceska, Jennifer. Phone interview, 10 June 2020.

Coffey, Emily. Interview, 24 February 2020, Atlanta Botanical Garden.

Earnhardt, Tom (host and producer). "Native Intelligence." *Exploring North Carolina*, season 1100, episode 1102 (2015). https://video.unctv.org/video /exploring-north-carolina-native-intelligence/.

Frances, Anne. Phone interview, 24 July 2020.

Moffett, Mincy. Phone interview, 22 July 2020.

Reese, Carol. Email communication, 17 November 2020.

Rush, Elizabeth. Public lecture during "A Planet in Balance: A Week Partnership with National Geographic Society," Chautauqua Institution, 19 July 2019.

Stewart, Beth. Email communication, 14 May 2020.

Further Reading

Carson, Rachel. *Silent Spring*. 1962. Repr., New York: HMH Books, 2002.

Davis, Donald E. *Southern United States: An Environmental History*. Santa Barbara: ABC-CLIO, 2006.

Kimmerer, Robin Wall. *Braiding Sweetgrass: Indigenous Wisdom, Scientific Knowledge, and the Teachings of Plants*. Minneapolis: Milkweed Editions, 2013.

Kolbert, Elizabeth. *The Sixth Extinction: An Unnatural History*. New York: Picador, Henry Holt and Company, 2014.

Pimm, Stuart L., and Peter H. Raven. "The Fate of the World's Plants." *Trends in Ecology and Evolution* 32, no. 5 (2017): 317–20.

Rogers, Kara. *The Quiet Extinction: Stories of North America's Rare and Threatened Plants*. Tucson: University of Arizona Press, 2015.

Weakley, Alan S. *Flora of the Southeastern United States*. Chapel Hill: University of North Carolina Herbarium, 2020. https://ncbg.unc.edu /research/unc-herbarium/flora-request/.

Index

Page numbers in italics refer to photos and captions.

Blomquist Garden of Native Plants at Duke Gardens, 208; mini-prairie habitat at, 208–11

Bloodworth, Stefan, 45, 222

Bloom, Sean, 190, 192

Blowing Rock, 9–11, *10*, 14–15, 17–18, 21, 25–26

Blue Ridge Mountains, 9, 223

Bluffs gayfeather (*Liatris gholsonii*), 38

bobwhite quail, 92, 201–2; habitat of, 206

bog cheeto (*Polygala lutea*), 65

botanizing, early techniques and tools, 9, 11–12

botany: decline in study of, 13, 71–72, 217–18; enhancing representation of women, people of color, and low-income students in, 69

bottlebrush grass (*Elymus hystrix*), 126

Braun, E. Lucy, 198

Brendel, Frederick, 13

Broad River, 12, 28, 99, 121, 139–41, 151, 153, 156, 172–73

Brooklyn Botanic Garden, 21

Buck, Lisa, 108–12, 115, 223

buttonbush (*Cephalanthus occidentalis*), 127

Byrd, Chuck, 54–55, 57–58, *59*, 60–62, 64–65, 67

Cahaba lily. *See* shoals spider lily

Cahaba Lily Festival, 108–9, 223

Cahaba River, 60, 84, 99–116, 118–20, 216, 223

Cahaba River Society (CRS), 109, 113, 115–16, 118–19, 216, 218

Callaway, Elvy Edison, 28–30, 35, 48, 95

Candeias, Matt, 51

canebrake, 158–59, 161

canebrake rattler, 65, 159

cane cutters, 167

cane Jakes, 167

Cannon, Brandi, 208, *209*

carnivorous plants, 2, 49–50, 199; and

black market, 51–52; pharmacological uses of, 64

Carolina parakeet, 77, 159

Carolina trefoil (*Acmispon helleri*), 149

Carson, Rachel, 5

Carter, Jimmy, 171

Castanea (journal), 41, 133

Catawba Land Conservancy, 180–81, 190–91

Catawba rhododendron (*Rhododendron catawbiense*), 14

Catawba River, 14, 99–103, 109, 157, 179–80, 183–87, 190

Catling, Paul M., 76–77, 93

Ceska, Jennifer, 147, 150–51, 217

challenge of protecting the planet, 218–21

Chapman, Alvan Wentworth, 79–81, 198; *Flora of the Southeastern United States*, 80

Chapman rhododendron (*Rhodendron champmanii*), 89

Chattooga Town, Chattooga Conservancy, 168

Cherokee Preservation Foundation, 164

Civilian Conservation Corps (CCC), 66

Clary, Renee, 16

Clean Water Act, 175

Clemson University, 145, 169

climate change, 6–7, 30, 43–44, 131, 133, 169, 203, 215, 217, 221

Coffey, Clarence, 200, 202

Coffey, Emily, 33, 40, 43, 67, 69, *70*, 151, 216

Colville, Frederick Vernon, 86–87

commercial development, 5, 89, 178

conservation, 1–2, 7–8, 211–14; ethics of, 212–13; and easements, 220; and genetic diversity, 33; and habitat destruction, 89, 199, 203; and habitat recovery, 31, 159, 218; and loss of grasslands, 203, 208, 210; and reintroducing species, 208; and safeguarding, 150

tulip tree (*Liriodendron tulipifera*), 24, 102

tundra swans, 103

University of North Carolina Herbarium, 20, 22

University of Tennessee, 196

US Department of Agriculture, 86, 88, 90

US Fish and Wildlife Service (FWS) 3, 9, 42–43, 51, 53, 56, 89–91, 95, 106, 110–12, 114, 168, 180, 206, 208, 219

Uzzell, Lila, 31, 47–48

Venus flytrap, 52, 64, 140

virtual field trips, 216

Wandersee, James, 16

warblers: Bachman's, 159; prothonotary, 102; Swainson's, 159

Wateree River, 157

water-quality violations and South River, 169–70

Weakley, Alan, 19–21, 22, 26, 180, 195

weaving with river cane, 162–66

White, Peter, 212–13, 222

white space, nature as, 118–120, 171. *See also* African Americans and nature

wild blueberries (*Vaccinium augustifolium*), 62, 205

wilderness, 6–7, 14, 18, 97, 127, 205

wildfires, 2

wild Florida azalea (*Rhododendron austrinum*), 80

Wildlife Restoration Act, also Pittman-Robertson Act, 146

wild strawberries (*Fragaria virginiana*), 15

William and Lynda Steere Herbarium, 24

Williams, Charlie, 178

Williams, Pandra, 153–55

Wilson, E. O., 1–3, 6

Wilson, Woodrow, 168, 190

woodlands, loss of, 5

Yadkin River, 10–21, 23–26, 149, 199

Yadkin River goldenrod (*Solidago plumosa*), 10–21, *19*, 23–26, 149

Yawn, Noah, 52–54, 56, 65, 67, 71, 72, 73, 218